To John Hatt, who brought me to Cumbria, challenged
everything that I believed about politics in the country,
inspired many of these letters, and who made me a better writer
and a better politician, and became both mentor and friend.

Middleland

ALSO BY RORY STEWART

The Places In Between

Occupational Hazards: My Time Governing in Iraq

The Marches: Border Walks with My Father

Politics On the Edge: A Memoir from Within

Middleland

Dispatches from the Borders

RORY STEWART

JONATHAN CAPE
LONDON

5 7 9 10 8 6 4

Jonathan Cape, an imprint of Vintage, is part of the
Penguin Random House group of companies

Vintage, Penguin Random House UK, One Embassy Gardens,
8 Viaduct Gardens, London SW11 7BW

penguin.co.uk/vintage
global.penguinrandomhouse.com

First published by Jonathan Cape in 2025

Typeset in 12.8/16pt Dante MT Std by Six Red Marbles UK, Thetford, Norfolk
Printed and bound in Great Britain by Clays Ltd, Elcograf S.p.A.

The authorised representative in the EEA is Penguin Random House Ireland,
Morrison Chambers, 32 Nassau Street, Dublin D02 YH68

A CIP catalogue record for this book is available from the British Library

ISBN 9781787336247

Penguin Random House is committed to a sustainable future
for our business, our readers and our planet. This book is made
from Forest Stewardship Council® certified paper.

Contents

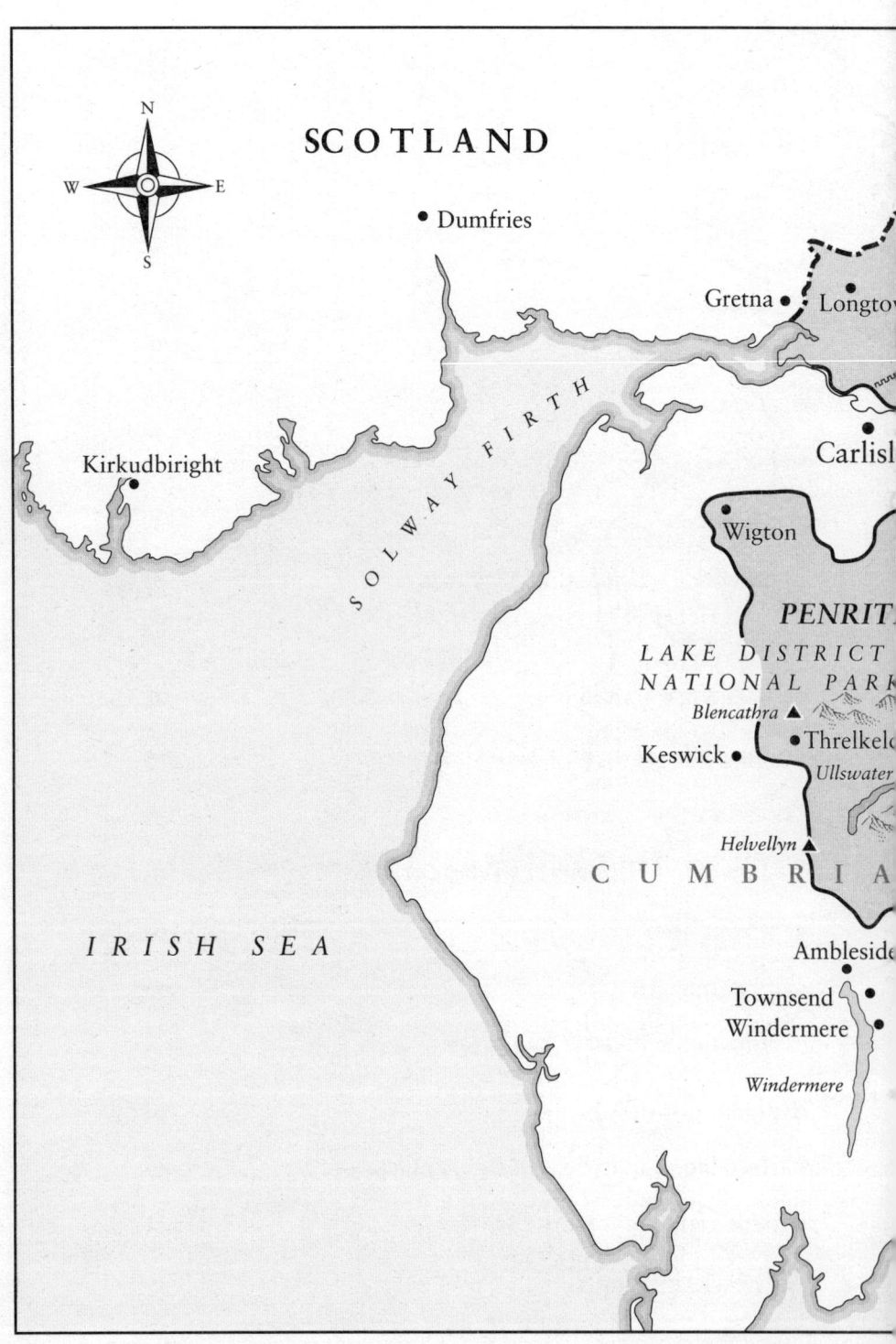

SCOTLAND

● Dumfries

Gretna ● ● Longto~

Kirkudbiright ●

Carlisl

SOLWAY FIRTH

● Wigton

PENRIT.

LAKE DISTRICT
NATIONAL PARK

Blencathra ▲

● Threlkel

Keswick ● *Ullswater*

Helvellyn ▲

I R I S H S E A

C U M B R I A

Ambleside ●

Townsend ●
Windermere ●

Windermere

NORTHUMBERLAND

TYNE & WEAR

Kielder Forest

• Bewcastle

Hadrian's Wall

• Brampton

North

Eden

Alston

• Lazonby

ND THE BORDER

DURHAM

Pennines

• Penrith
• Eamont Bridge

Askham
Rory's cottage
• Bampton
• Shap
weswater

• Appleby-in-Westmorland

Eden

• Brough

Kirkby Stephen •

• Kendal

NORTH YORKSHIRE

Historical Timeline

AD Pre-70 (Iron Age)   **State of the Carvetii**

The parliamentary constituency of Penrith and the Border was, during the Iron Age, an independent state: the nation of the Carvetii, which probably means the deer people. One of their important shrines seems to have been at Bewcastle and was dedicated to a god called Cocidius, who looks – in surviving depictions – a bit like a beetle with a spear and shield.

Its material culture seems to be quite distinct from (and poorer than) those of the tribal kingdoms of southern England or northern Scotland.

AD 70–410 (Rome)   **Military frontier**

The Romans invade Britain and capture the state of the Carvetii. They run Hadrian's Wall through the state, cutting off its northern areas and leaving the shrine of Cocidius marooned seven miles north of the wall.

They turn the state into a Roman civitas or tribal territory. It has its own local senate, but the Romans station an astonishing concentration of troops from across the empire in the territory and seem to have governed much of the modern constituency under direct, brutal military rule.

410–c. 1000 (Post-Roman)   **Brittonic kingdom(s) –**
Rheged / Strathclyde /
Cumbria

After almost 350 years, the Romans leave. In their place, and sometimes in their forts, a new local warrior aristocracy emerges. Penrith and the Border again becomes the centre of an independent kingdom: a Middleland stretching at times as far north as Glasgow.

Its linguistic culture is related to modern Welsh. It has its own dynasty of kings and bards and epics and saints. Perhaps its most famous ruler is King Urien. The kingdom is then called Rheged.

In later periods it is called Cumbria or Cumbraland – which means, in Welsh, simply 'our nation'. It comes under pressure from the Anglo-Saxon kingdom of Northumbria in the east, and later from Norse settlers who come in from the west.

But as late as the tenth century, the Cumbrian king is still meeting on almost equal terms with the kings of England and Scotland.

c. 1000–1300   **Forgotten border and royal forest**

By the eleventh century, the kingdom has been squeezed out of existence between the expanding realms of England and Scotland, and it has become a debated no man's land. The Scottish kings still crown their heirs 'Prince of Cumbria'; the land is not included in William the Conqueror's Domesday Book.

Half of the constituency is turned into a royal hunting reserve and the people are largely excluded from much of the territory.

A great deal of the remaining land is then given to cross-border monasteries, originally from France, who begin to drive agricultural and economic development.

By the late thirteenth century, great border lords such as the Norman family of Robert the Bruce have lands and interests in both English Cumbria and the now Scottish parts of Cumbria.

The borders are indistinct and cross-border landholdings are common. The people of Cumbria continue to be recognised as having their own identity and customs in law books.

1300–1603 (Border Wars)   **Western March**

At the end of the thirteenth century, the war of succession to the Scottish throne turns the whole area into a frontier war zone. What is now Penrith and the Border ultimately becomes the English Western March, governed by a Warden of the March. The northern part of the old independent nation becomes the Scottish Western March. In between lies a no man's land called the Debatable Lands.

For the next three hundred years, the constituency is one of the most dangerous areas in Britain – a zone of proxy wars and border raids by 'reiving clans' – governed by the Warden of the Western March under 'border law'.

The war, however, paradoxically preserves the cultural connection across the old Middleland – the clothes, ballads, laws and practices of borderers form a single culture quite different from England or Scotland.

1603–late 1700s   Middle Shires and Romantic discovery

When the Scottish King James VI is also crowned King of England as James I of England, the war between England and Scotland ceases. He renames the area the 'Middle Shires'.

Paradoxically, peace intensifies the legal and administrative differences between English Cumbria and the part of the old Middleland now in modern Scotland.

By the late eighteenth century, the area has become the cradle of the Romantic movement. Walter Scott celebrates the ballads and history of the borders; Wordsworth, Coleridge and Turner, the landscape and culture of the Lakes.

1800–2010   Extraction and early industry

Mining for iron begins in Cumbria in the early modern period, but by the time of the nineteenth-century industrial revolution, west Cumbria is increasingly dominated by coal mining. The fell farms of Penrith and the Border seem increasingly poor and marginal by comparison.

In the twentieth century, the remote location and sparse population of the constituency attracts munitions factories and missile test sites.

By the 1950s, farming, and particularly small upland sheep farms, remains the cultural heart of the constituency. But tourism has become the largest income earner.

By 2010, Penrith and the Border is the largest and most sparsely populated constituency in England, the only constituency with 'border' in its name, and contains half the English–Scottish border. It is a safe Conservative seat. It is in England, but it is coincidentally represented exclusively by MPs from Scottish families, of which the longest serving and most celebrated is Willie Whitelaw, Margaret Thatcher's deputy prime minister.

Preface

I wrote these pieces between 2009 and 2019: as the liberal order of the early twenty-first century collapsed into its shadow form. I began soon after the financial crisis of 2008, at the moment when Iraq and Afghanistan were fully exposed as 3-trillion-dollar catastrophes, and when Twitter and Facebook were shattering the old media. I wrote through Xi Jinping's rise to power, ISIS' creation of a caliphate and the flight of 5 million Syrians; as Putin invaded Crimea; when a Brexit referendum was lost, and Donald Trump was elected for the first time.

But the 'dispatches' in this book are about none of those events. They were written for the *Cumberland and Westmorland Herald*: a newspaper, founded in 1860 and addressed to the inhabitants of the Eden valley, on the edge of the Lake District in the far north-west corner of England. The paper included letters from angry voters. But it was better known for its detailed accounts of agricultural fairs and livestock prices; its coverage of charities, deaths and snowstorms; and its old photographs from primary schools, printed for the benefit of people whose grandchildren were now attending their old schools. My column appeared, usually every two weeks, under the letters. It was called 'The Word from Westminster', because I was the Member of Parliament.

That is a problem. My other books have been written in retrospect and at a distance. These are pieces of journalism, written at the time, by a politician addressing local electors. My former colleagues in the House of Commons will wince. And with good reason. Some of these pieces, which I think of as letters, have the breezy optimism of a headteacher's report or, even worse, a parish notice. Re-reading them I was reminded of Reverend Yorick reviewing his sermons, in Laurence Sterne's *Tristram Shandy*:

—I don't like it at all . . . This is but a flimsy kind of a composition; what was in my head when I made it?
—For this sermon I shall be hanged – for I have stolen the greatest part of it.

And I was tempted to follow Yorick's example and hide this collection in an old magazine cover, stinking of horse drugs. Instead of which I have got rid of two-thirds, fixed some grammar and punctuation, left in my self-serving comments and shallow analysis, and published them. Why?

Because I think the letters show something about local politics on the ground, at a time when 'Politics' seems ever more inflated, global, virtual and melodramatic. They reveal some of the radical strangeness of our democracy: one man, trying to understand and then represent the separate needs, desires and dreams of 100,000. They record my flaws and prejudices as an MP. They illustrate how government ignored, frustrated and sustained lives in unpredictable combinations. And they are an elegy to a very unusual place. A constituency that had roots stretching back hundreds of years, but which does not exist any more.

Much, of course, was amiss. This constituency, like other parts of the country, had front gardens dense with nettles and broken toys; and fetid waterways lined with Japanese knotweed and syringes. Knocking on the door of a Georgian house, I found a stumbling man in his eighties, whose grand dining room was littered with used and stinking nappies. The door of a smaller house was opened by a younger man, too distraught to speak, and too terrified, it seemed, to step outside. Such suffering could not be skated over with happy talk about 'joined-up adult social care' and 'well-being'. There were businesspeople who bullied, lied and polluted. (One tried to bribe me.) There were 'free-range' egg farms, in which the chickens never left the sheds in their six short weeks of life and were only stopped from wounding or eating each other, in this crammed mess of stinking feathers, by having their beaks painfully trimmed with hot blades or lasers. The Eden valley was not paradise.

But it was not simply political propaganda when I said that I had never lived anywhere so improbably beautiful. It was true. The intimate, evolving encounter of Cumbrians with the natural world was preserved in the thousand miles of hand-built dry-stone walls – granite, sandstone and limestone, glittering with quartz and mica – which separated meadow, pasture and fellside into separate patches of coloured altitude. I walked through every one of the 120 villages in the constituency and developed a deep admiration for Cumbrians – even if I found it difficult to record my admiration without sounding fey, obsequious or antique.

I remember, for example, a tall man who tapped on my car window in a snowstorm, having been up since two in the morning, searching for a one-year-old sheep on a mountainside. I am tempted to write about his 'dignity and courage'. But this doesn't quite capture his thin ribcage, or large cracked hands, the precision of his footsteps in the snow, or his rather lengthy dismissal of my heated car – which was, simultaneously, an insult, a joke and a world view. Nor have I fully captured my friend from Mallerstang in these letters. She was hardly over five feet tall, with a past that included being a barmaid and a pony trek organiser, and unlike the farmer in the snowstorm, she did not seem like a Viking saga hero. But she forced the government to subsidise housing, broadband and footpaths, in remote places where officials insisted such things could not happen. She embodied 'hope': but only insofar as hope meant being innocently sceptical, slyly flattering and courteously blunt.

It was no accident, perhaps, that the constituency had once been an independent country. The upper frontier of this ancient nation of Cumbria had moved back and forth over time, sometimes stretching almost as far as Glasgow, and sometimes fading back into the sodden clay of the wastes of Bewcastle. Often, however, the state's boundaries coincided almost exactly with Penrith and the Border. The featureless moor where the eastern edge of my constituency met County Durham was recognised as a national frontier by centuries of Northumbrian, English and Scottish kings: it was here that Eric Bloodaxe, the last Viking ruler of York, was slain.

The place-names, stone walls and even bloodlines in Cumbria still demonstrated startling continuity with these ancestral identities. The dry-stone walls of Yanwath, first built in the late Bronze Age, were maintained for livestock by a modern community. The largest town in the area, Penrith – a word from the lost language of the kingdom of Cumbria probably meaning 'Red Hill' – still had the highest concentration of Viking DNA in the UK. Many of these letters, however, are also about startling, extreme change: about empty places, which had once been densely populated, and tranquil valleys, which had for centuries been industrial mines.

In modern times, we were also defined by our sparsity – in

this most thinly populated constituency, the parish of Bewcastle had one-fifth the number of people that it had had 1,800 years earlier, while in the same period the population of England had increased forty-fold. The Eden valley had a population density one-twentieth of that of the south-east of England. Our sparse population encouraged government officials to treat us almost as an empty space in which to dump munitions factories and missile launch pads, commercial forestry, reservoirs, military ranges, nuclear power and nuclear waste. Later still, others treated us as a new kind of blank space: a zone onto which they could project fantasies, not of war, but of rewilding, aiming to turn the slopes again into what they had not been for at least 6,000 years – unfarmed wilderness.

But all the regulations, subsidies and impositions from Westminster, more obvious in a rural area than almost anywhere else, never seemed to undermine a dense web of voluntary organisations, and a powerfully self-reliant culture. The constituency had among the lowest average incomes in the country but also had one of the lowest unemployment rates, and was consistently recorded in surveys as one of the happiest places in Britain.

I had ways of communicating with constituents, other than these articles. I often received and replied to more than 20,000 constituent emails in a year. I met many thousands face-to-face. I also spent too much time – sometimes two hours a day – on my addiction to Twitter. But I probably found writing these articles more difficult than anything else. The deadline for the *Herald* column was Friday morning. Each article took me six or seven hours, with the first three hours spent writing 800 words, and the remaining hours spent deleting and rewriting.

I have excluded two-thirds of the pieces which I wrote from

this collection, because they were too local, too international or too dull. But I have included five more personal letters from the *Herald* (they appear as 'interludes'). I have also kept some material which has appeared in my other books on this period, although I have tried to minimise the overlap.

I did most of the writing after I had finished my day's work on Thursday. My wife Shoshana often generously read drafts back to me. And hearing them in her voice transformed them – revealing where the articles had become too political, pretentious or muddled. But when I did not start writing till six, Shoshana was asleep long before I had got the article into a state to be read. On those nights, I wrote till one in the morning, or sometimes, three or four. Often, I got into the bath for the last two hours, propping my laptop on an adjoining chair, and trusting to a combination of cooling water and terror at dropping the computer to keep me awake through the editing process.

I finished many articles in the kitchen. As I wrote, I could hear the scufflings of the stoat that had its burrow beneath the counter, and sometimes the high-pitched screams of its prey. I could sense, beneath my feet, the stream which had once run along the kitchen floor, as a convenience, but which now only re-emerged in heavy rain. The gas stove scalded my calves, filled the kitchen with methane, and my head with fumes.

Why did I find these letters such a struggle to write? It was partly because I was trying to describe an area and people which fit none of our standard assumptions about nationhood, identity or economic life, and which demanded a quite different vocabulary. My constituency was the imaginative soul of Britain. This was the place where Keir Starmer's parents took him every year from the London commuter belt, and where David Cameron did his wild swimming. But our population was too small, entrenched and dispersed to deliver the sort of dramatic national change which leaders promise before elections. Government policy required generalisations, statistics and abstractions, which my constituents did not supply. The narratives of neoliberals or their opponents, nationalists, populists or even centrists seemed to have little relevance in Penrith and the Border, and my voters often made concepts like 'growth', 'democracy' and 'product- ivity' seem comically absurd. They felt forgotten by politicians. And they often were.

But re-reading the letters reminds me that much of this is also true for most of Britain outside the south-east of England, and, of course, for much of rural America. Penrith and the Border was an extreme example – the largest, most sparsely populated constituency in England – but its unique- ness brought a clarity and focus to more universal themes. This constituency, which often seemed, on the surface, to preserve the world of the past, also suggested what other democracies might become.

So much for the grander themes. In the end, my strong- est regret is that – partly because I did not want to seem too romantic – I failed to express how intensely I loved the place. It was less embarrassing to record the glint of an abandoned shopping trolley sunk in river mud than to describe the wonder which I felt on Knipe Scar, hearing the long call of a thrush, catching the sunlight on the broad face of Bampton church, or meditating within the scarlet sandstone walls of St Ninian's Cave. It was more comfortable to joke about Colin's bulk on his tiny quad bike, or Steve's welded rust, than to say how much I admired the wild strength with which they ploughed life out of heavy soil. Easier to tease, than to thank and praise the impres- sive and improbable people with whom I spent the most stimu- lating decade of my life.

OUR LAND

PENRITH AND THE BORDER WAS, until its abolition by the Boundary Commission, a constituency in the far north-west of England. Its northern edge was formed by the English–Scottish border, of which it contained the western half. When I was elected to the constituency – as a Conservative MP – in 2010, my southern frontier was formed by a range of hills called the Howgills, and in the west by the sharp volcanic ridges of the Lake District mountains of Helvellyn and Blencathra. Its eastern frontier was where Cumbria ended on the bleak, round tops of the Pennines.

At the centre of this horseshoe of granite and volcanic slopes lay a valley called Eden. The free-draining soil on the sandstone supported large herds of dairy cattle. The Pennines, and the Lake District valleys in the constituency, running up from Haweswater and Ullswater, supported upland sheep farms. Beyond Hadrian's Wall, four hundred people lived in the wastes of Bewcastle on a hundred square kilometres of wet clay soil. And beyond that lay a stretch of country, long disputed between England and Scotland, and often rejected by both, called the 'Debatable Lands'.

The largest town was Penrith, with a population of 14,000 people. Seven other settlements were recognised as towns: the smallest, Brough, had 750 residents; the largest, Wigton, had 5,000. The others were Longtown, Brampton, Alston, Appleby-in-Westmorland and Kirkby Stephen. Two-thirds of my constituents lived outside these towns, scattered across a hundred villages, two hundred hamlets and thousands of isolated buildings. Butterwick, the hamlet in which I lived, consisted of eight houses. But I was a long way from my neighbours, half a mile up a steep farm track, through two gates on a limestone crag, backing onto open fell.

As the pieces in this first chapter suggest, it was difficult for outsiders or officials to understand such a place. Nor, as an MP, was it simple to 'represent'. The first piece was written six weeks after I was first elected as the MP for Penrith and the Border in 2010. In the long walk that took place two years later, every new person I met exposed differences in lives and attitudes that defied my attempts to categorise or generalise. Steve Pattinson, whose farm I describe visiting here, worked the wet, bleak expanses beyond Hadrian's Wall. He often put me up, invited me on his initiatives (I joined him for example on a Young Farmers' bicycle pub crawl in 2010) and in turn supported mine (he joined me on horseback in 2019 in my attempt to create a new border riding festival).

Newton Rigg, which is described in the third essay, was the only agricultural college in the area. Established a century earlier with generous gifts of land by local farmers, it had a strong tradition in upland farming and dairy. Like so many Cumbrian institutions – from the community hospitals to the community ambulance at Alston – its income was struggling, and it was on the verge of closure.

6 December 2014

I approached Steve's farm up the wet slope from the river, kicking the clay off my boots. Steve was in his yellow-and-blue boiler suit, stacking a trailer, and his ten-year-old son was examining a pet calf. Steve had been up milking at four in the morning, and it was now five in the afternoon, but he stacked quickly and energetically. It was easy to believe that he planned to cycle ten miles before returning to milking that evening. He saw me, stopped what he was doing, and smiled broadly. With his round glasses perched on his nose, and his grin, he almost seemed younger than his son.

His daughter cooked for us. After tea, he took me down the valley. The first fields we passed contained bedraggled and scrawny sheep among glossy spikes of rush. The field belonged to another farmer. 'It's just sheep ground. It's just enough for sheep,' said Steve.

'Enough?'

'I'd say comfortable rather than thriving. They're hill-bred sheep, they're Swales in there. They can live in it but not much else could.'

The surrounding land had a thick clay base. Most years, the rain wrecked the fields: pools of dark stagnant water gathered on the chocolate-brown mud; the cattle destroyed the grass roots and the deep structure of the soil.

'Do any of these roads get cleared of snow?' I asked.

'No.'

'How do you get the kids to school?'

'In the winter it's down to us with 4x4s. Two years ago when we hit minus seventeen, I was taking the pick-up down. The kids had to finish school at dinner time and get up the road and spread the grit.'

Back in the farmhouse, Steve put me up in the main downstairs room, in front of an open fire. I slept right through the night, woke to find that I had left my wet clothes scattered everywhere, pulled them on, and went outside. Steve was in the milking parlour, a tiny stone shed in the corner of the farmyard. The equipment was forty years old, and he was having to learn to weld rust. He had only sixty cows. It was not easy farming.

The supermarkets and the milk processors were driving down the price of milk (it was in real terms half of what it had been twenty years earlier). He had joined a farmers' co-operative, which had then gone bankrupt. The new buyers were charging high haulage fees.

Trevor and Colin came into the yard – neighbouring farmers, his age, and friends. Colin was on his quad bike, with a child sitting on the handlebars. He was six feet four inches and very broad. He dwarfed the quad, looking as though he was riding a child's push-car. His hair was blonder, his cheeks even pinker than Steve's.

A hundred yards from Steve's farm stood the old school. I had passed the old church, the old shop, the old petrol pump, the old village pub and the old auction mart. All seven community institutions had been abandoned. Only the farms remained. But they, too, were beginning to disappear.

The Forestry Commission had begun to plant more of the grazing with conifers; the edge of Colin's land had been forcibly converted into a peat bog conservation zone; competition from farms, ten times their size, in the lowlands was forcing neighbours to sell for holiday homes. We can only estimate the population of Bewcastle when it was a Roman fort, but it seems to have been 2,000 – which was then one for every 2,000 people in England. In 1841, 1,274 people had lived in the parish. Now, there were only 411, while the population of England was twenty times larger than in Roman times.

But Steve, Colin and Trevor were not giving up. It was their land, a place, they said, where you could 'feel free', where people left you alone. Steve pointed to a patch that he thought had promise. 'You could get that back into production. Personally, I would graze hard, and clean the ditches out, and then start spreading muck on some of the good places, then I'd get out on a weed wiper and get rid of a lot of these rushes; a little bit of fertiliser on some of the better places which burns a lot of the rushes away . . .'

Steve had collected reconditioned computers for farmers, he had helped run IT training courses and built – 'after a bit of a battle' – a website for the local school. He had experimented with better breeds of cattle and won the Highland Show with his vast muscled British Blue bull, whom he had named 'Colin' in honour of the size of his friend.

Trevor and Colin had taken major steps with draining, and new grass types. Year after year, the brutal weather destroyed what they had achieved with the soil. Year after year they tried again.

Their small family-owned farms had been on almost exactly this footprint since the fifteenth century. But it was the people more than the landscape that mattered: rare reservoirs of vitality. Steve's ten-year-old son passed me as I left, solemnly taking his calf out of the yard. He said that he was thinking of taking over his father's farm.

18 August 2012

It is the parliamentary recess, and I have been walking through Cumbria and the Borders. On the second day I climbed over Helvellyn and Great Dodd, and slept in cloud on the top of Blencathra. I found when I had cooked my porridge that I had forgotten to pack a spoon to eat it with. Day six was along the sand from Maryport to Silloth. Day nine was over lowland raised mires from Wigton to Bowness. Fording the Solway, the waist-deep water was silver. In the country between Longtown and Bewcastle, the rain-gorged rivers were sometimes chocolate with clay, and sometimes black with peat. Days fifteen and sixteen were spent in the 200,000-acre spruce forest of Kershope and Kielder.

There seems to be more time for conversations when walking, and for encounters which would be unlikely in a car or a chair. In Wigton, I had breakfast with community workers, coffee with apprentice engineers in the factory canteen, dropped into the youth club, looked at a new park, lunched with the manager of the auction mart, called on a town councillor and the rector at the church, joined a patrol of the community policing team, had supper in the kebab shop, and finally slept on a sofa belonging to a mother on the Greenacres housing estate.

Connections stretched tightly across the county: the farmer who gave me coffee by Dearham was the brother-in-law of the farmer who gave me coffee at Plumpton. I saw Emerald the Limousin cow at John Elliot's farm, having seen her earlier, seventy-three miles away, at Matt Ridley's. (She had been moved by truck to visit one of John's bulls.)

At Kirkbride the soil was peat-dark, the drainage ditches choked. Bowness Common was thick with sphagnum and bog asphodel. In the Bailey valley, boots sank in heavy clay and, despite the new draining and planting, the soil was poached by the cattle, and the blue water oozed over a dozen 200-acre owner-occupied farms. Across the Kershope Burn, however, the farms were tenanted at 2,000 acres, and the countryside was empty of homesteads: the legacy of the Scottish Border clearances. Each half-day brought a different geology, or altitude, or rainfall, or landholding pattern – and different problems stemming from different soils. When I walked between the two related Saxon churches of Bridekirk and Dearham, I transitioned from the limestone fells to the coal seam, and found at Dearham an explosion of hedges, nettle, meadowsweet and deprivation: in an hour's walk, the average life expectancy had fallen by a decade.

In the nineteenth century scholars recorded oral traditions in every valley: the ballads about the Captain of Bewcastle, and stories about the King of Patterdale – a trace of a memory of a time when this had been an independent kingdom. Each hill had its story. No longer. The 'characters' who were remembered in Longtown's Graham Arms were still alive in the 1980s.

Trevor Telford had learned that his ancestors had lived on almost every farm in that upper valley over the four hundred years since his reiving ancestors flowed south, but he had discovered this from parish records; his grandparents had not told him this, and probably had not known. The man who could name sixty plants in the hedgerow had moved here from Leicestershire. It was not in the perfect medieval villages but in the new housing estates that everyone seemed to know and be related to their neighbours and be in and out of each other's homes. My host in the 1950s estate in Wigton knew the inhabitants of a hundred houses, and could name all their children and aunts; I doubted that the IT consultants in their seventeenth-century grange knew more than two people in their village.

Within this landscape, these boundaries, and this history I encountered a bewildering diversity of professions and experiences. I forded the Solway with a publican who was also a *haaf-netter*, fishing in that 1,200-year-old Norse tradition. I climbed Helvellyn with a man who was a potter, a solar panel installer, a beer festival organiser and a web designer. The Silloth harbour-master who walked with me from Allonby had trained in Dubai, and introduced me to the captain in the harbour, who was from Kaliningrad. The manager responsible for Carr's flour was the sixth generation of his family to be a miller, but he was operating brand-new machinery spouting the purest flour through ever-juddering steel tubes. Eric Weir was knowledgeable not just about his Swaledale flock, but also about the history of lead mining in the Howtown hills. Barry Todhunter, the huntsman of the Blencathra fox hunt, had a name which implied his ancestors had been fox-hunters in the Middle Ages, but his passion was riding the American railway system.

There is nothing straightforward or orderly about such conversations: the eccentricity, the learning, the charm and often the bluntness of a hundred meetings on footpaths. Ours is a Middleland of frontiers, and of encounters along the ancient Marches of England and Scotland. Such stories and lives could exist nowhere except in Cumbria and on the Scottish border, nor could they have flowered in just this way, even here, a decade ago.

Twenty-three days into the walk with another twelve to go, it feels as though the energy of our place springs from neither history nor landscape, but from incongruous lives and unexpected voices.

16 October 2010

On Friday morning I discussed the role of an MP with a roomful of eleven- and twelve-year-olds. A girl from Patterdale suggested I was there to improve rural services. A boy from Skelton focused on the cost of fuel. They explained the difficulty of making a profit in a village shop. And they had strong ideas about the constitution. 'Your job is to advise the prime minister,' suggested a boy from Stainton. 'Your boss is the prime minister: it's his money you're spending,' suggested a girl. But when I asked, 'Who are really my bosses?' they turned this on its head. 'We are,' they replied. 'What would happen if I just sat around drinking coffee?' 'We'd vote you out.' 'Whose money am I really spending?' 'Ours.'

If I'm lucky enough to still be around in seven years' time, I would enjoy representing the next generation of voters.

But perhaps I could have explained an MP's role better not with theory but by showing them what I do hour by hour. I did this for a student from Kirkby Stephen Grammar School. He came with me from the primary-school meeting to Penrith hospital to hear about breast screening in Carlisle and the new GP-commissioning process; then to Eamont Bridge to look at the new flood protection barriers; then to a room above a curry house in Penrith for East Cumbria Family Support's annual meeting, and then to the gleaming glass pentagon that housed the Environment Agency. He was not able to attend the morning meeting on Newton Rigg agricultural college, see the fourteen people who came to the surgeries that afternoon at the George Hotel, or come to the speech at Longtown memorial hall that evening. But he sat through discussions on clinical procedures and survey data; tender and procurement processes for child-care; and mathematical modelling of river flows.

I wonder what the primary-school children would have made of this day. What would they have thought of the meeting on the future of Newton Rigg, for example? It was, I think, the fiftieth meeting I have had on Newton Rigg with academics, farmers and ministers in Cumbria, Cheshire and London. We have been joined by staff from the county council, and a group from the district council. A senior scientist and a Member of the European Parliament turned Conservative lord have written a report. A former Liberal Democrat councillor has come up with a slightly different plan – and there is a third plan from the staff of the college, who have offered to take a pay-cut. I have been making freedom of information requests on their behalf, and asking friends to examine accounts in Carlisle and Lancaster to establish whether Newton Rigg is really losing £4 million a year, as the last vice-chancellor implied.

We all agree that we do not want a forced marriage, yoking Newton Rigg to an institution which does not 'get it'. But it is going to struggle on its own. Could we match the college with a partner who wants the relationship, is wealthy enough to underwrite the deficit, understands land-based education and isn't going to asset-strip and leave?

What kind of role has a Member of Parliament in all this? A cross between an intelligence officer – trying to ferret out the information that the Yorkshire bidder for Newton Rigg and the civil servants would prefer to conceal – and a marriage counsellor, trying to bring untrusting parties together?

And how would you explain the other fifteen meetings on Friday? The East Cumbria Family Support meeting was about the detail of government policies on charities; Eamont Bridge was assessing engineering work. In a meeting with Sainsbury's, I was like a pedantic lawyer, pushing for a promise that the access to the old town centre from their car park would be 'direct', 'immediate' and 'pedestrian'. In broadband meetings, I was raising money.

Would the children have seen a pattern in all these roles? What expectations do they have on how an MP should use his time? It is easy to lurch between optimism and pessimism: to think politicians could construct a perfect state or do nothing at all. But my sense is that politics starts from an awareness of limits. A sense of what politicians don't know. Most of the time I am not initiating or deciding but instead doing that ambiguous thing called representing – which I think could also be described as juggling incompatible roles, managing my own ignorance, choosing between lesser evils, and reinforcing energies which are already there without me.

FARMLAND

F ARMING WAS THE HEARTWOOD of the constituency, which contained 2,000 farms, tens of thousands of cows and 2 million sheep. These farms had average incomes of £8,000 a year. This was a shocking figure, even taking into account any creative agricultural accounting. It was far below the minimum wage. Fifteen years before my birth, almost all the houses outside the towns and larger villages lacked indoor lavatories and electricity. Since then, power and sanitation had arrived, but at least a third of my constituents still lacked access to broadband or a mobile signal. When the snow closed the pass, it could take two hours or more to get an ambulance to Alston.

The farmers were responsible for the very colour of the Lake District and Pennine hills. On the lower slopes, they kept dense flocks of sheep, who mowed a short lawn, rich in brilliant green chlorophyll. Beyond the last walls of the *ring-garth* was their *out-bye*, where the hardier sheep were hefted on thin acidic soils, at perhaps one-twentieth of the density. On this land, generally owned in common, Herdwick and Rough Fell picked their way through stiff ageing heather, sprung bushes of blaeberry, and sere clumps of pale sheep's-fescue and purple moor-grass. It was these patterns of land ownership and stocking, reinforced by the dry-stone walls, which created the startling stripes of colour that transfixed visitors – the bright emerald-green lawn on their

lower slopes, and the purple, russet and fawn on the wild upper slopes.

The land, which the farmers owned and worked, was tightly constrained by the regulations and restrictions of unelected public bodies, including Natural England, the Environment Agency and the three national parks. Basic traditional farming practices – burying fallen stock on the land, for example – were now prohibited. Despite the lack of affordable housing, development of any kind was highly restricted. Many of the valleys were being rewilded under the influence of large national charities such as the National Trust and Royal Society for the Protection of Birds – losing sheep, farmers and even the colour of the hillsides.

Many of the original letters in the *Herald* were about farming. Often, I agreed with farmers. But sometimes I didn't – I angered farmers, for example, by pushing in one of the letters, not included here, for compulsory vaccination of cattle against bovine TB. It was, as I knew from helping my father manage his Highland cows, dangerous and time-consuming to inject cattle; but I felt that if we didn't do it, we would lose our herds. I failed to implement the policy. And it looks like I was wrong. Fifteen years later there is still no compulsory vaccination, and Cumbria still has very few cases of bovine TB.

Much more serious for the survival of farming, however, were the bigger global issues of climate change, global trade, European subsidies and markets. Extreme floods which were traditionally considered 'once in a hundred year' events struck farms three times in the eight years between 2007 and 2015. Farmers already found it very difficult to compete with imported Australian or New Zealand sheep, or to survive the lurches in the prices of milk – which often fell below the cost of production. Ninety per cent of the farms would have been bankrupt without European Union subsidies. But these subsidies came with regulations on everything from field margins to stocking density, which made it increasingly difficult to run a viable business.

Due to these and other factors, the number of farms was falling fast. At the time of my father's birth in 1922, there were still many farms in Cumbria of thirty acres or less. By the time I became the MP, my friend Steve's sixty-acre farm was considered tiny – below the viable minimum. Larger units were now well over a thousand acres. Two-thirds of farming families had vanished over a few decades. As the farms disappeared, the basic structure of rural life was broken: farmhouses became expensive second homes in empty valleys. We lacked the farmers' children who kept local schools open. We lost the farmers' work, their memories, their long intimacy with the land. We lost the heritage which gave much of the life, and interest, to Cumbria, for *off-comers* as much as locals.

Central government had little sympathy for any of this. There was still a department called the Department of Environment, Food and Rural Affairs, in which I eventually became a minister. But my boss, Liz Truss, the Secretary of State, said that she saw no valid distinction between urban and rural areas, and there were only four civil servants left in the 'rural affairs' part of my portfolio. As for the other two nouns in the department title – 'food' seemed to mean large industrial farms in the south of England, and 'environment' increasingly seemed to mean rewilding nature by excluding upland farmers and their stock from the fells.

Yet Cumbrian farmers, who were battling restrictions and regulations, and working long hours in rough weather for very low incomes, never saw themselves as victims. Most of them wanted their children to take over their farms. And, although some of the children wanted to escape, there were more young farmers who wanted land than could find it. Their attachment to soil and flock was extreme. Their predecessors, centuries earlier, had hefted their sheep on the high fells, which meant flocks had been taught to stay on a particular slope, with no need for walls – and this tight, fierce connection to a quadrant of soil was passed by ewes on to their lambs. Britain has more breeds of sheep than the rest of the world put together. In Cumbria, different breeds specialise in different moorland soils. The most famous are the Herdwick – sheep with grey fleeces, white woolly faces, and lips which seem drawn into a placid smile – who have been known to survive for days buried in snow.

16 February 2013

On Sunday I stood with a friend, looking at his fields, which slope down from the fellside to the water. The river was in spate, and silver lakes had formed in the meadows. He had got up early to save his Swaledales from drowning. But where was he to put them?

His drier fields, to the north, had rare herbs, and he had agreed with Natural England not to put more sheep on them. And he owned no other fields. When he began, he had thirty acres, like others in the valley. Now he has a hundred, but that counts as a small farm today.

A bad year, like this one, with poor sheep prices, torrential rain and waterlogged ground can be catastrophic. He guessed that within a generation there would be only two farms left in the whole valley – large ranches, with hired labour – and all the remaining farmhouses would become homes for the wealthy. His sheep were dying, 'many because of fluke, but others, I feel, are just giving up'. As I walked up the fellside, I passed another dead sheep sprawled across a yellow-ant tussock.

On Monday, I tried to represent this stark and fatal erosion of rural life in Parliament and to press for more funding for the rural schools, transport and housing. I was joined by a dozen colleagues from other rural seats. Many were former local councillors, and they showed a detailed knowledge of the concepts of the moment – from 'discrepancies in formula grants' to 'damping'. They demonstrated that the grants we received ignore the cost of heating rural homes and overlook our low wages and minimal public transport. People in rural areas received 50 per cent less per head than people in urban areas, paid more council tax and received fewer services.

But the ministers sat shaking their heads at every plea. When one of the ministers replied, he said he'd just heard the same argument made for urban areas. And he had his own statistics. He appeared to feel that it was all just special pleading.

So how do you convince someone in an urban area to care about rural communities? We tried arguments on food security, and the future demand for food (the protein that is fed on our wet grass will become more valuable as the world population grows, and wealthy Indians and Chinese want to eat more meat). We argued that tourism is one of the most rapidly growing sectors in the British economy, and that rural areas are the key to it – people come to Cumbria not to see a picturesque wilderness, but to be part of a historical human landscape and community. But such economic arguments could always be answered with another blizzard of statistics, suggesting we could make money in a different way.

What we were failing to convey – or the ministers were failing to grasp – was that living rural communities are part of most people's idea of Britain; they are part of the imagination even of people in cities. We are one of the most advanced developed countries in the world. It is more than two hundred years since we began to clear our countryside of people, and build mills, factories, tenements and civic palaces across green fields. And yet by a miracle, our rural areas still hold some traces of where we all began.

This precious survival is exceedingly fragile. Our organic living communities, our young families, small farms, remaining schools and men, like my friend, are under extraordinary pressure from a dozen directions. They are struggling with environmental regulations, rising fuel costs and the powerful market forces which undercut all small rural shops, services and farms, day after day. (Who can stand up to a supermarket?) Without support – what our opponents define as 'subsidies' – the character, culture and history of our rural areas will be lost forever. And we would become a countryside composed only of wealthy people who have moved here to escape the cities.

Cumbria is not the slopes of Spain, where the hillside is scarred and packed with half-completed concrete time-shares. Nor is it yet New Hampshire, where you can walk from one holiday cottage to the next through tight, scrubby secondary woodland, aware only when you stumble across a half-broken dry-stone wall that there were once any small farms there at all.

It has not even yet become the Scottish Borders: that landscape of 2,000-acre farms, whose tenants are rarely from local families, and where the sheep in the heather are untended, and field after field has not a human in sight.

Cumbria remains, by some astonishing combination of soil, luck and cultural stubbornness, still just enough of a living human landscape, rooted and connected through local families to a local past: the landscape many of us would like our grandchildren to see.

Somewhere in the welter of statistics, we need to remember that one of the great responsibilities of government is to preserve some of what we cherish. And that includes our small farms and our living villages. There is no civilisation in a wilderness for millionaires.

22 June 2013

All the government policies and subsidies on farming, and all the messages of every English political party, for twenty years, seem to boil down only to 'food production' or 'the environment'. (I deliberately exclude the Scots, the Welsh and the Northern Irish who have a more thoughtful understanding of farming.) These two slogans define Westminster's understanding of all the rules and subsidies which emanate from the 'two pillars' of the multi-billion-pound Common Agricultural Policy of the European Union. In the name of these two objectives, England seems bent on copying the landscape of the United States, seeking to replicate either the monoculture of Kansas or the wilderness of Alaska: a binary vision in which land becomes either an industrial factory for the production of the maximum food at the cheapest price, or a national park almost devoid of human cultivation.

Yet policy-makers do all they can to pretend they have 'nothing against small farms'. They even occasionally produce ideas for helping them. Their standard suggestion is that a well-developed marketing strategy might provide 'a premium for specialist, local produce'; or that a flexible and innovative co-op could increase the leverage for small farms. But, in practice, organic producers – supplying Neal's Yard in the West End of

London – or Danish-style farmer cooperatives are immensely difficult to sustain.

One reason is obvious. If you have sixty cows, a single farmer is tied to them every day of the week, twice a day. But twenty times as many cows might only need five times as many people, and people are expensive: you can produce more milk, more cheaply. Scale allows you to build up the reserves to ride out fluctuations (sheep or feed prices can halve or double in a year), allows you to diversify and increases your leverage with a buyer.

Then there is the pressure from environmental schemes. Biodiversity and carbon-capture targets require that wetlands and mires increase, that mosses and ferns re-emerge, that there is ground cover, that peat and tree cover expands to sequester carbon. Protecting biodiversity and carbon sequestration means fencing some areas entirely off from stock, reducing the overall number of sheep per hectare (sometimes to a quarter of what the land once held, sometimes to a twentieth), and reversing the draining, cutting, grazing and moorland management on which larger flocks depend.

When environmental regulations force a sheep farmer to reduce his flock by three-quarters, the income he makes from selling sheep falls by approximately three-quarters. Initially, governments made up this shortfall with environmental subsidies, which amounted to roughly the amount the farmers had lost from their farming income (this was called 'income foregone'). But this was a one-way street. As flocks diminished, fewer people became shepherds, and as the moorland management ceased much of the pasture was degraded. If the environmental subsidies ceased, and the farmers tried to restore their income by bringing their land back into production, they would lack the skills and capital resources (particularly for investments like installing field drains) to do so. They would have little alternative other than to leave much of it as a semi-wilderness and go bankrupt.

The environmentalist George Monbiot suggests that this is indeed the plan – that the subsidies were only ever intended to be kept in place long enough to erode farming practices, and that they would then be abandoned, leaving the farmers with no flocks, no crop and no income. In his words, 'Subsidies in their current form surely cannot last much longer . . . farming will gradually withdraw from the hills . . . rewilding (the reintroduction of wolves, lynx, wolverines) . . . is better deployed in the uplands.'

The enterprises that are best able to survive the demands both of the markets and the environmentalists are the large estates, including some of the oldest aristocratic holdings. They have the scale and wealth to ranch sheep on vast hillsides, and ride out the fluctuations in the market. They have the resources, the staff, the time, and the agents to put together the most complex and ambitious proposals for extra environmental subsidies.

Officials do not acknowledge what is happening. Instead, we still hear reassuring sounds about the upland farmers' role in creating and maintaining the network of dry-stone walls, the barns, the environment and 'rural services'. But I also hear MPs from all parties and many officials mutter, 'these family farms are too small, their land is too marginal, it is inevitable that they will fold', while the environmentalists quietly encourage farmland to be handed directly to the RSPB, or planted with trees, and the National Trust allows water to ruin the lowland pastures of their small tenant farms, apparently on the advice of the Environment Agency.

It doesn't have to be like this. France and Japan put policies in place to support small farms, because of their intrinsic value to society. Britain may not be ready to go that far, but we could at least begin to count the number of farms we are losing, and ask officials to reveal what impact their rules and policies and subsidies are having on small farms.

We could ask charities like the National Trust to make the interests of small tenant farmers a priority equal to their environmental targets. After all, Beatrix Potter gave her farms to the National Trust to preserve the traditional farming culture and the rare sheep breeds – not to eliminate them. Our requirement is not for more agricultural subsidies, but instead to target them on preserving small upland farms rather than on wolverines or vast estates.

And we would all benefit. Farms are not only mechanisms to maximise profit, or to support non-human species. Small farms are also bearers of culture: Lake District valleys are the legacy of more than a thousand years of cultivation.

Their presence is what makes the British countryside still feel quite different from that of the United States. Our small family farms are a final fragile link to our past. We will miss them terribly when they are gone.

12 November 2011

On Friday, in a Cumbrian conference centre, I was looking at a long sandy beach, below a tropical hillside, in the sun. And on my left, Stephen, who lives on that hillside, was talking about banana yields. Listening to the speech were two hundred people, including: Robert, who keeps a dairy herd on the slopes of the Pennines, a suckler cow farmer from Armathwaite, a miller, a service station owner, councillors, Lake District sheep farmers and the chief executive of a large London charity. There was a reason why so many Cumbrians were there.

St Lucia is a sliver of windward land in the Caribbean, facing the Atlantic, with a patchwork of small family farms scattered across a rugged landscape, beneath a central mountain as high as Helvellyn. It is a remote place, far from markets, whose economy – like Penrith and the Border's – is based on tourism and agriculture. And it was recently on the verge of economic extinction.

From 1991 to 2007, the value of St Lucia's banana trade fell from $71 million to $16 million, and the number of farmers (in a country with a population a quarter of Cumbria's) plummeted from 10,000 to 1,200. An industry which had made up one-third of the GDP of the country in the 1980s almost ceased to exist.

Thousands of small family farms which were the core of

rural communities and the heart of the country; the farmers who created the landscape which the tourists came to see; their incomes which supported the rural schools, clinics and shops; the culture which gave pride and self-respect to St Lucians – all were being destroyed.

Many assumed that it was inevitable: simply supply and demand – cheaper bananas were grown elsewhere, and if St Lucia couldn't compete, it should just do something else. But the Fairtrade movement fought back. They persuaded customers to pay more for St Lucian bananas under a Fairtrade label; they passed on a fair price to farmers, who in turn improved standards on farms, and put surplus income into community projects from bus shelters to clinics. Stephen feels they saved his farm.

So, what about Cumbria, which suffers from the same global commodity prices and the same ruthless retailers? Small farms are also our lifeblood, created through centuries of effort, sustained through a tough, intelligent approach to the soil. They sustain families, who pay for the rural economy, manage the environment and support tourism. And as the world population soars past 7 billion, and 100 million more middle-class Asians emerge every year, there is an ever-growing demand for food – and in particular the milk and meat in which Cumbria specialises. Saving our farms could, therefore, also be a sensible long-term commercial investment.

But in my valley, we are in danger of losing the land, the skills, the stock and the people, to the point where we could struggle to rebuild again. Milk prices (driven down by supermarkets) are barely at the break-even point, and the number of dairy farms in the UK has fallen from 28,000 to 11,000 in only fifteen years.

Could something more be done to save Cumbrian

farms – through a new brand, or cooperatives, or other ways of securing fair, predictable prices?

My own experience makes me cautious. I once started a charity to save traditional Afghan crafts: beautiful woodwork, ceramic tiles and jewellery, in which Afghans took justifiable pride. We hoped through saving the crafts to support jobs and communities. But the results were mixed. We created an institute to retrain craftspeople. We invested in machinery and improved design and quality. We sold the crafts abroad: persuading a London hotel to build a suite from carved Afghan wood, building three hundred beds for the US Embassy, exhibiting calligraphy in Bahrain, getting Afghan jewellery into Fifth Avenue stores and into the British Museum's shop. It was tough. At times, we were able to make enough money to support a primary school and clinic in Kabul. But a reliable market was difficult to establish.

Customers recognised the quality, but they were often not prepared to pay enough to cover the craftsman's time and the transport costs, let alone the cost of training the craftsmen. Afghan jewellery has done reasonably well but we still struggle for commissions to support woodcarvers. Ceramics is a disaster. All the skill and dedication of the craftsmen and craftswomen, all their contribution to their communities, are still undermined by less distinctive, but cheaper Chinese goods.

Fairtrade faces similar challenges. They have to persuade independent-minded farmers to cooperate with each other and trust them. They have to convince customers to trust their brand and pay more for it. They have to persuade the major supermarkets to stock their products (and anyone who has tried to sell direct to a major supermarket – as Bells of Lazonby, for example, does in Cumbria – can testify to how challenging that is). They have to maintain these fragile long-distance relationships and markets for years.

Fighting for fair terms and fair prices against the blind, grinding, short-term pressures of the global markets is brutal. It requires extraordinary goodwill, energy, imagination, knowledge, drive and luck.

Nevertheless, Rheged, on Friday, suggests that it can be done. It is possible to get fairer and more reliable prices for farmers. And the communities and places in which they live can be transformed if you succeed.

In the decade since I wrote many of these letters the situation has become far worse. First, Brexit took away the European Union's Single Farm Payments; then Boris Johnson's trade deals opened the area to imports of cheaper foreign livestock; and finally, the Keir Starmer government changed the taxes on farm inheritance and closed the environmental subsidy schemes. Small parts of this may be reversed but, fundamentally, English politicians and civil servants – unlike their French, Swiss, Norwegian or Japanese equivalents – seem to lack a commitment to the tradition and heritage of small family farms. The catastrophic fall in livestock and the number of farms is accelerating, leaving more and more of the landscape unpastured, unpeopled and, in important ways, unrecognisable.

FORGOTTEN LAND

P ENRITH AND THE BORDER WAS the centre of what had been for many centuries an independent kingdom of Cumbrians – older than, and quite separate from, the kingdoms of the Scots and English. Scholars keep discovering from archaeology, place-name analysis, comparative history and DNA that early Cumbrians were often immigrants – radically alien in their beliefs, customs, languages and values.

When our tiny state was first recorded in writing, by the Romans, it was known as the 'civitas' or state of the 'Carvetii'. Other areas of Britain at the time of the Roman invasion had coins, well-made pottery, delicate metalwork and impressive settlements. The Carvetii did not. They lived in a scattering of small family farms, and their archaeological traces often amounted to little more than plough lines, just visible on dry soil in a winter sun.

But it was here – not in the wealthier, more developed areas – that the Roman invaders concentrated their military occupation. The whole of Egypt, with its literacy, bureaucracy, architecture, vast population, fertile fields and religious marvels, was peaceably held by 5,000 Roman soldiers. This tiny stretch of thin, wet, cold soil and sparse population drew in at least 10,000 – perhaps one soldier for every two members of the local population – and the Roman Empire, which stretched in a great uniform parade of roads and aqueducts, columns and capitals, from Tunis to Cologne and Baghdad to York, ground to a sudden halt in this almost invisible culture.

After the Romans left, this mini-state of the Carvetii re-emerged as an independent kingdom – or perhaps from the Roman perspective, reverted to tribalism. It had its own language, kings, bards and saints. It survived Vikings and Northumbrians until it was squeezed between the expanding kingdoms of England and Scotland. Its last independent king died not long before the Battle of Hastings in 1066. Then the expanding kingdoms of England and Scotland tried to crush it out of existence and turn much of it into a royal forest – the medieval equivalent of a national park (although the punishment for troubling the wildlife was the death penalty). Or it was given over to monasteries.

During the English–Scottish wars from 1304 to 1604, it became again what it had been during the Roman occupation a thousand years earlier: a frontier zone, under military rule. Farmhouses were turned into stone towers, churches into forts, houses into castles. Secretive agents governed the constituency not by English or Scottish law but by a 'border' law, which legitimised much older tribal customs – such as seizing a clan member in punishment for a relative's crime. The governors, or 'Wardens of the Western March', often behaved as though they were punishing foreign tribes rather than serving their own citizens. But the new frontier with Scotland, like Hadrian's Wall before it (half of the wall lay in the constituency), instead of fracturing the culture, seemed to bring people more tightly together. And when the fighting ended, the astonishing beauty of the area made it the crucible of European romanticism: the landscape of Wordsworth, Coleridge and Turner.

I love the fact that Cumbria had been a Middleland, sitting between England and Scotland, challenging the borders and identities on which those larger states wanted to insist so stridently. Its status as an independent nation shook the cosy British idea that our nation state was somehow self-evident and natural. Every outside power had tried to redraw Cumbria – Roman security, Norman hunting, Victorian industry, twentieth-century armaments, modern ecology – leaving ghost-layers for the next plan to stumble over. The perennial cycle of erasure, occupation by outsiders, rewriting and amnesia foreshadowed what I found in Cumbria in 2010. This chapter develops these themes.

20 July 2013

The Elizabethan traveller William Camden, who visited Cumbria more than four hundred years ago, didn't leave a very satisfying picture of our local history. He was not helped by the fact that he was visiting at the end of a 300-year border war, and found the people around Hadrian's Wall to be 'ranke' and dangerous. The first detailed studies of our parishes were conducted by the Victorians – often clergymen, who filled the time between their Sunday sermons analysing Cumbrian dialect, sketching Dark Age monuments and transcribing the records of abandoned monasteries.

But they and their academic successors still saw Cumbria in the Iron Age as little more than a remote county of England. They thought that the valley floor around the rivers Eamont and Eden, at the time of the Roman conquest, was uninhabited marsh and forest land (with farms only on the higher, lighter limestone soils). They described it, after the Roman departure, as part of the Anglo-Saxon kingdom of Northumbria, and then incorporated into England. Through all of this, Cumbria rarely seemed anything more than a side-show of a side-show.

But if all that were true, why did five kings meet 1,100 years ago at Eamont Bridge? When you are driving up the A6 or down the A66, it hardly seems a likely place for a royal conference.

But that was where the monk who wrote the Worcester Chronicle recorded a meeting in the summer of AD 926. King Athelstan of England held a meeting 'aet eamotum', at Eamont, with the kings of Scotland, of Wales, of Northumbria. And – most crucially – another monastic chronicler informs us that the fifth king present was the king of *Cumbria*.

The background to this astonishing story has only been fully explored by recent academics. It remains largely unknown to the rest of us, because it is hidden in articles behind the paywalls of academic journals, or in horrendously expensive books not available in local libraries. Most of us continue to get our sense of Cumbria by surfing between amateur conspiracies on Arthur and Merlin, and the dry, unexpansive gobbets about parish populations embedded in Wikipedia.

If you can access the recent research, however, you will find it has reversed almost everything that we learned in school. First, recent historians have established that Cumbria was not part of the tribe in York, it was its own pre-Roman tribal territory: an autonomous state, separate from those across the Pennines or the Solway. Our nation – as confirmed by a tombstone and a milestone near Eamont Bridge – had a name, 'the Carvetii'. Digging at Yanwath has made it clear that settlement wasn't only on the uplands. Instead, on the valley floor was a network of dry-stone walls, arable fields and stock stretching over three miles, and behind it at Clifton, a seven-acre site impressive enough to have been a royal capital.

In the mid-1980s, two professors published a 120-page analysis of the farming, industry, communications and trade of this tiny Iron Age community, whose boundaries were approximately those of Penrith and the Border. Then in 2005, another academic proposed that there were not one but two Iron Age tribes in our area: the Carvetii of Cumberland were pro-Roman, the unnamed tribe of Westmorland, south of Eamont Bridge, were anti-Roman, while the Romans were based by the bridge to man the frontier.

The recent archaeology around Eamont Bridge has also suggested what happened when the Romans left. At first, one is struck by a sudden absence in the archaeological record – no coins, no formal pottery, not a single stone building is found in the next 250 years – and the uplands were abandoned, returning to peat bog and scrub. The Romans, who might have assumed that after three hundred years British society would be transformed permanently in their image, would have found that their material civilisation had vanished.

But aerial photography has now shown that on the riverbank by Ninekirk, under the legend of St Ninian, lay an extensive set of ditches and rectangular buildings, suggesting a Dark Age monastery. The Anglian cross found near my cottage at Lowther might mark both (as at Bewcastle) a border and (as at Hexham) a religious community. And we can fill out this picture of our lost nation with epics written in the ancient Cumbrian language and mentions in the chronicles of neighbouring kingdoms.

Recent research is showing ever more clearly that Cumbria, rather than being a subset of its neighbours, was a unique and sometimes powerful actor, independent of what we now call England and Scotland, for almost seven hundred years.

But our local schools do not teach enough about our local history. Instead, the National Curriculum encourages pupils to plod dutifully through the story of Anglo-Saxon England and King Alfred – which has almost nothing to do with us. I would like us, instead, in twenty years' time, when waiting for the lights to change at Eamont Bridge, to remember that this hamlet was an international frontier, and the meeting place for an international conference with our own nation and our own king.

7 January 2012

You could once ride from modern Syria through Bulgaria and the Netherlands, on fine roads, using a single language, in a single state, until, halfway between Brampton and Longtown, Rome stopped. And 2,000 years later, it is still true that just beyond Longtown, one nation – England – stops, and another, Scotland, begins. This is the only constituency with 'border' in its name. Through the Middle Ages, Cumbrians held our land in a 'border tenancy', were ruled by a 'border law', and were ravaged by a border war.

Even today, the frontiers exist in more than motorway signs. When I gave a talk in Penton, I was speaking to farmers from a five-mile radius, and yet you could identify every Scot in the room, because two hundred yards across the border the accent changed completely.

Yet for seven hundred years after the Romans left, what is now the border was not a border. It lay at the centre of a single Middleland, half in modern England, half in modern Scotland, independent of both, and belonging to neither. As late as the eleventh century, we were not in King William the Conqueror of England's Domesday Book, and our patron saints – Mungo (also known as Kentigern), even Ninian – were not English saints. But nor were we part of Scotland either, despite the pretensions of King David (who mined his silver in Alston).

We are the heirs of this independent kingdom, which had sometimes been called Rheged, Strathclyde, the old North, Yr Hen Ogledd, but always, probably, Cumbria. ('Cumbria' in our ancient language simply means 'our people' or 'our nation'.) A nation with its own language, spoken long before the Roman conquest, before the seaborne Irish Gaelic-speaking invaders of Scotland, or the seaborne Anglo-Saxon-speaking invaders of England, or the Norse-speaking settlers in the Lakes. We had our own line of kings, with their bards and genealogists: so confident that long after the Romans left, they still led their warriors into the Highlands, to Tyneside and to the edge of Wales. So legitimate that even when this kingdom had been reduced to a narrow stretch between Carlisle and the Clyde, Owen of Cumbria was still treated as a brother-king, alongside Athelstan of England and Constantine of Scotland.

But stand in the central lobby in Parliament and you will see the arms and saints of England, Scotland, Wales and Ireland, but of our old kingdom, the Middleland, no trace. Where did the nation of Cumbria go? Why did we not reappear? The United Nations has many artificial confections, some of which we have invented and invaded, from Iraq to Libya. The scattered islands of South East Asia, Christian and Hindu, Animist and Muslim, have been reconfigured as Indonesia. Beside the Adriatic Sea, even mountainous Montenegro has become a nation. But Cumbria, independent for longer, larger and more populous, has not.

Our neighbour, Alex Salmond, is concocting full independence for his nation, partly through denying an older Cumbrian identity. Highland bagpipes and Highland kilts are imposed on the lowlands that had nothing to do with either. So, too, between Edinburgh and Dumfries, the Scottish nationalists have placed road signs in a language that was never spoken in our area. They say: 'Failte gu Alba.' This is a jingle in a dialect of Irish Gaelic placed in our Middleland, which spoke Cumbric, a relative of modern Welsh, and which would have said instead: 'Croeso y Cumru.'

No nation, not even our ancient nation of Cumbria, is inevitable and eternal. Like the Philippines or El Salvador, Cumbria grew out of an artificial colony: the backbone of our kingdom was that extravagant exercise in imperial megalomania called Hadrian's Wall, which planted fourteen forts, eighty castles, 240 towers, and subsidised and paid tens of thousands of men for three hundred years. Like every nation, too, our nation was not a single people but a land of immigrants of many faiths: the Romano-British Cumbrians mixed with descendants of the Anglians who built the cross at Bewcastle, the Norse worshippers of Odin at Oddendale, and even perhaps the Syrian archers whose god lay at Kirkby Thore. Its borders moved too. What was once called Cumbria probably did not include the portion of modern 'Cumbria' to the south of this constituency, and certainly included Scottish territory in what is now labelled Dumfries and Clydeside.

But there was still something more than lines on a map which sustained our kingdom for seven hundred years and made it the last Welsh-British-speaking part of England. When the ancient Britons had been driven from everywhere else, something drew them here; still counting their sheep in the old language, in places still named in Cumbric: 'Lyvennet', 'Blencathra', 'Penrith'. Some aspect of our sea-fringed moors and fells, of our city – the culture of King Urien of Rheged, Urien 'city-born', 'Urien Y Eochydd', 'Lord of the Rip-tide' – then and now, gave us an identity quite distinct, and almost national.

Willie Whitelaw first stood to be MP in Penrith and the Border against William Brownrigg. Some of Mr Brownrigg's programme seems a little anachronistic: fair wages for mole-catchers, the reintroduction of cock-fighting, no clipping the tails of Clydesdales. Mr Brownrigg did not recognise Cumbria's traditional territories of Dumfries and Westmorland, and he received only 368 votes. But he half sensed something in us, which never existed in Wiltshire or Norfolk, when he demanded, in his 1955 election address, 'Home Rule for Cumberland.'

18 September 2012

This afternoon, I saw the piece of silver which has just been uncovered by a metal detector in a field by Crosby Garrett. It is a Roman cavalry mask, and it is 2,000 years old. The top was found in more than thirty pieces. The patina on the bronze back is like old green cracked leather. But the shining silver front is a perfect image of a human face. The small lips are slightly parted. The face is a little fat with a soft chin. The mask is topped by a cap from which his thick unruly curls escape. The face looks very young, little more than sixteen, but very serious. One wide eye is focused on the scene in front, but the other seems to look into another world. It is to my eyes an image more of a priest than a warrior.

This mask would have been worn for cavalry exercises: perhaps near where it was found in Crosby Garrett. Its owner would have been a Roman officer. A man one social rank below the senators but still far wealthier than the average citizen. He would have needed to be to buy this mask.

At first I thought it odd that such an expensive luxury had been brought 2,000 miles to the frontier and wondered whether it was an eccentric exception. But a hundred years ago, another mask was excavated in the Scottish Borders and two hundred years ago, one was found in Lancashire. The Roman writer Arrian describes their use:

> The horsemen enter [the parade ground] fully armed, and those of high rank or superior in horsemanship wear gilded helmets of iron or bronze to draw the attention of the spectators. Unlike the helmets made for active service, these do not cover the head and cheeks only but are made to fit all round the faces of the riders with apertures for the eyes . . . From the helmets hang yellow plumes, a matter of décor as much as utility. As the horses move forward, the slightest breeze adds to the beauty of these plumes. Instead of breastplates the horsemen wear close-fitting Cimmerian tunics embroidered with scarlet, red or blue and other colours. On their legs they wear tight trousers, not loosely fitting like those of the Parthians and Armenians.

The other masks which have been found in northern England and southern Scotland have flat tops but the Cumbrian Crosby Garrett helmet has a great, foot-high rearing Persian-Armenian cap, topped by a griffin. The others look fierce and impersonal. But this one is troublingly human.

When the other Roman cavalrymen cantered onto the field to join the display in their own shining metal masks, they would have noticed this one. Were they impressed? Or would they have thought the wearer was showing off? Much must have depended on who wore it.

It is tempting to assume that he looked like his mask: a very young, perhaps slightly over-protected son of Rome sent out with this fancy expensive gear on his first trip to the frontier. But why should the wearer resemble his mask? He might have been a wizened veteran commanding the fort at Brough; or serving with the cavalry by Carlisle.

Why is it topped by a cap from what is now the Iranian frontier? Was this a reference to an Eastern god favoured by the Roman military? Or – given that there were Syrian archers then at Kirkby Thore – was the owner himself claiming links to the Middle East? Was it just fashion? Or a joke?

And what does all this incredible expense and style, these bizarre costumes for ritualistic horse manoeuvres, suggest about life not far from Crosby Garrett 2,000 years ago? And what made him leave this very expensive thing behind? Who folded it and placed it face-down in the Cumbrian mud? And, most importantly, how are we going to try to keep it for Cumbria?

We put together a bid to keep the helmet in Cumbria. It combined a local museum with the British Museum, but on 7 October 2010, we were outbid at Christie's. The helmet was sold for £2.3 million (US$3.6 million) to an anonymous private buyer.

15 March 2014

The central section of Hadrian's Wall – which cuts through the north of our constituency – is one of the most remote landscapes in Britain. It is a place of scattered, isolated farms, held by the same families, some perhaps since the time of the Vikings.

There is no mobile signal and few buses, let alone superfast broadband. The 'global world' feels very far away. There is almost no immigration. And it is easy to imagine that the area was even more isolated and traditional in the past: an ancient, subsistence farming economy, with little movement, little sign of man or government.

But 1,800 years ago, central Cumbria and Northumbria was densely populated, filled with immigrants, drawn from 2 million square miles of Europe and Asia. The land was marked by a succession of vast, contemporary structures, built to a standard specification developed in the Mediterranean. It was powered by a continual flow of manufactured goods from factories in what is now France and Germany. The population fed on food imported from as far away as North Africa.

A ceaseless circulation of bureaucrats generated hundreds of thousands of documents, which moved back and forth along the very latest communication systems. All this is buried within the landscape that we live in today.

When I walked from Newcastle along Hadrian's Wall, I passed the intersection of the two main Roman roads – now a large roundabout with a pub and petrol station – and entered land which was rougher and apparently more sour. Crops vanished, replaced first with dairy cows, then with suckler cows and sheep. I followed a section of the wall that ran three stones high over the wet, green fellside, across becks and through gullies.

At one point, dozens of horned Swaledale ewes charged past me, scrabbling over the parapet of the wall, pursued by a furious sheepdog. The collie drove the flock back out of the barbarian lands again into Roman territory, and the sheep leapt and shoved. One, panicking, somersaulted over the stone borderline. An old farmer stood at a distance on a ridge, controlling the dog. He did not pull his crook out of the soil. The wall ran straight through the middle of his pasture, and was no longer even a field boundary. He looked as though he had been in the landscape forever.

The low line of stones at Birdoswald, on the other side of the gorge from Gilsland, now houses only a single caretaker at night. But eight hundred years ago, what was now an open field containing a dozen Simmental cows had been the barrack lines of a vast military camp, occupied by a thousand Dacians. These men were immigrants who came to Birdoswald from a land 3,000 miles away.

They tended to be long-limbed redheads with startlingly blue eyes, whose leaders had once worn baggy trousers and floppy felt hats. Their homeland stretched from modern Romania to the Black Sea: a place for growing fine figs and olives. Their great capital at home included a sundial twenty feet long, capable of advanced chronological calculations. At home their annual rainfall was ten inches. They would have struggled to make their sundial work in the Cumbrian rain.

Just as Highland regiments carried bagpipes, so the Dacians carried their ethnic dragon standard, the Draco, which howled when the wind passed through. Just as Highland regiments in the eighteenth century were allowed to wear kilts after the kilt was banned in Scotland, so the Dacian regiments were allowed to carry their curved weapon in Britain, after it had been banned in Dacia.

They would have brought their own cuisine (just as a Roman African regiment in Scotland carried a distinctive casserole dish). They venerated their king, Decibalus, who killed himself with his fighting sickle, rather than submit to Rome. (You can still see a portrait of him doing so, on Trajan's column.)

In the fort of Birdoswald, carved in the Cumbrian sandstone, is a perfect image of the Dacian fighting sickle, to commemor-ate their building of the granary between AD 205 and 208; and a tomb of a child, named Decibalus. He had been given the name at least a century after the last Dacian King Decibalus had killed himself. These were apparently people who believed in tradition.

Later, I clambered over a fence to inspect a deep, broad ditch, with high grass mounds on either side, running straight across a field. A farmer raced up on his quad bike. His family had owned the land on either side of the wall for more generations than he knew. 'Hundreds of years, I'd say.' His cows were moving up and down the slopes of the ditch, searching for fresher pasture among rushes and thistles.

'What is it?' he asked. 'I heard someone say, it's an Iron Age fortification built long before the Romans.'

I said that it was the 'vallum' defence, built by the Romans at the same time as the wall.

His farmhouse was built from stone quarried by Roman

soldiers. You could still see the marks of their chisels, cutting diagonally across the square stones.

Between AD 120 and AD 410, when the farmer's ancestors may have been Vikings in Scandinavia, perhaps 12,000 Romanians lived, worked and worshipped in this now empty field. And their long limbs must still lie in the surrounding soil.

I found this piece on my hard drive when bringing this book together. I didn't send it to the Herald *for publication.*

The very first surviving written evidence from Penrith concerned a politician. The 1,800-year-old tombstone ran: 'Flavius Martius, senator and quaestor of the civitas of the Carvetii, of quaestorian rank, who lived 45 years.'

The civitas of the Carvetii was a Roman political district that roughly coincided with the boundaries of this constituency of Penrith and the Border. A local senator and quaestor meant something between a modern councillor and a modern Member of Parliament. Flavius was my age when he died.

We know nothing more about Flavius as an individual. But we know a very large amount about people like him. The cavalry officer whose beautiful silver mask has been uncovered recently in Crosby Garrett was almost certainly a wealthy Roman. But Flavius, despite his Roman name, was almost certainly not. Nor was he from the many ethnic groups, such as the Romanian Dacians, who were imported by Rome to man forts such as Birdoswald. He was a local Briton, drawn from a family of the Carvetii tribe, speaking Cumbrian – our ancient language related to modern Welsh. Before the Roman invasion, his ancestors would have tattooed their bodies, carried shields and swords decorated with sinuous serpentine metalwork, and worn torcs: a stiff necklace with a ball on each end, its metal band thick as a hazel staff. At some point, however, his family had decided to work with the Romans, and Flavius will have received a Roman education as well as a Roman name.

By the time Flavius went to school, the time of his tribal ancestors near Penrith had long passed. His family had exchanged cattle raids and boasts in smoke-filled wooden halls for the square plastered walls of a small Roman villa. His vision of political leadership would have been formed not by Celtic war bands but by reading Cicero in Latin schoolbooks. Roman textbooks taught that being a politician was not just one job among many, but the highest possible profession and activity known to man. Cicero insisted that politicians should understand everything from the army and the treasury to local laws and international treaties. Flavius was expected to 'excel in oratory, knowledge, hard work and memory' and to create a state that was prosperous, renowned and celebrated for its integrity. He would have learned from Aristotle that his vocation was about allowing citizens to achieve 'eudaimonia' – happiness in its very fullest sense. Politics was the route to a godlike fame: an 'art which brings into being merits of an extraordinary and almost superhuman character . . . the life by which a truly great man achieves greatness'.

But even 1,800 years ago, the representative for Penrith would have had to reconcile the philosopher's ideal of politics with the realities of his day job. His grand title and elevated public status might have encouraged his constituents to hold him responsible for many things. But he had little real power, and no control of central budgets. Many of the things that he might have wanted to do – changes that seemed obvious to him and his constituents – could be challenged or blocked by a lawyer. Roman law – sovereign, intricate, comprehensive and uniform – governed everything. (You can still see the local laws that governed a district like Penrith because they are preserved on six bronze tablets, each the size of a 43-inch TV screen, unearthed on a scrubby hillside in Spain.) He was not leading an independent tribe, he was representing a small, remote part of a super-state, in what we might call today a single market and customs union, underpinned by a single currency and a set of pan-European laws.

And when he wasn't tied up in rules and regulations, or carrying the can for a new set of taxes or cuts, he was constrained by the limits of the budget, by priorities set in a distant, uncomprehending capital, and by the network of informal pressures from the wealthy, the pious, the aggressive and the corrupt.

How distant all of this must have seemed from the personal power wielded by his tribal ancestors; how much at odds with the grand visions of classical heroes which dominated his schoolbooks. How different to the kind of personal charismatic power that a tribal leader might accrue if they could regain their independence from Rome. Did Flavius reconcile himself to the daily disappointments, the small humiliations, the public contempt? How much of this did his family acknowledge, when they chose to commemorate him on his tombstone, not as a father or a husband, but as a politician? As one of the 'honestiores' – the 'honourable member'. For Penrith.

1 October 2011

I wanted to spend some time with my father. He has always enjoyed walking, and talking about the army, Scotland and empire. So we decided to walk Hadrian's Wall, and I expected long discussions about the parallels between British and Roman colonialism. But I had lost my voice and he is eighty-nine: I could not talk much and he could not walk much. And he was less interested in the wall than I expected.

In the Great North Museum at Newcastle, he left me at the Roman tombstones, and sat and read a book on wild mushrooms. He claims that when I was six and he took me to the Great Wall of China, I said: 'Once you've seen one temple, you've seen them all.' That, he said, was his view of Roman camps. Eventually we left the wall and went home. Our best day together – we travelled by car – proved to be a visit to his childhood home, and the graves of his parents.

But for me the walk along the wall was an unsettling revelation. It is easy in Cumbria to feel a connection to our Norse and Anglo-Saxon past: we can worship in a Saxon church in Morland; my cottage follows a Viking floorplan; our dialect can be understood by a Dane; Norse words are part of our modern vocabulary; and there is, I imagine, Scandinavian blood in most of us. But the wall is the most dramatic reminder of our Celtic-Roman history. And it suggests things far more alien, extravagant and brutal than I had ever imagined.

I have heard historians describe the wall as 'a permeable trading post' and emphasise how much melding there was between the British and Roman populations. But at Wallsend, the excavations have revealed a line of fortification hundreds of yards wide: a ten-foot turf wall, followed by a twelve-foot ditch, followed by a berm set with spikes and thorns, then a fifteen-foot stone wall, then another ten-foot mound, another fifteen-foot vallum ditch and a ten-foot mound. These fortifications run almost unbroken for eighty miles. For me, they do not suggest a zone of gentle inter-cultural communication.

I once lived in a fortified camp in Al Amara in provincial Iraq with five hundred British soldiers, surrounded by a line of giant sandbags. The nearest neighbouring camp was in Basra, sixty miles away. But in the Roman wall, there was a manned tower every three hundred yards, a castle every mile, a fort with a garrison the size of ours in Al Amara every seven miles, and an additional line of large forts two miles south (as at Vindolanda and Corbridge), and other smaller outposts just north (as at Bewcastle). These were auxiliary positions. There were also three full legions in Britain – more than in any other comparable province of the Roman Empire. And the Romans held these positions not, like us in Amara, for three years, but for three hundred years.

There are some British details, but overwhelmingly the inscriptions, the clothes, the buildings, even the shoes found along the wall are relentlessly Roman. Not far from the wall, the British continued to live a life in roundhouses, similar to those that existed long before the Roman arrival. A Libyan could become an emperor but no Briton did; indeed, very few ethnic Britons were given jobs in the wider Roman Empire. Even the auxiliaries may not have been as integrated into British life as we imagine. The Syrian archers beyond Housesteads worshipped a Syrian god; the Batavians in Vindolanda were like Gurkhas – a separate ethnic military elite – and they have left notes referring contemptuously to the 'britu, nculli': the pathetic little Britons.

Why did Rome maintain this cripplingly expensive occupation? The smaller walls on the German, Saharan and Iraqi frontiers protected Rome from millions of people in Africa, Europe and Asia. But in this case, there was only a sparsely populated Scotland beyond. Britain never posed a serious threat to the Roman Empire; and it never brought in enough revenue to justify the expense of holding it.

I had some beautiful days walking. And I would love to work to reconstruct more of the wall, to ensure that we light the beacons annually, and to make a tighter connection between the walkers and the forts. But we should never pretend it is a comfortable history. Every step made me more conscious of an almost pointless military occupation, maintained largely because for Roman politicians, as for their many successors, 'failure was not an option'.

If Britain had really had the easy relationship with Rome which some imagine, more would have survived (as it did in France, for example). But when the legions left in AD 410, almost four hundred years of Roman civilisation collapsed overnight. Within a decade, from Cumbria to Kent, there was no coinage, the potteries and aqueducts had stopped, the villas had been abandoned, writing had largely been forgotten. And for us no trace remained except for some ditches to inconvenience the plough. This great symbol of the brutality, stubbornness and pride of empire was reduced to a stone quarry eighty miles long. For 1,500 years it was robbed for farmhouse, and barn, and dry-stone wall.

To add to the sense of an alien occupation, an excavation of a fort at the border in 2025 found that a quarter of the Roman shoes discovered there were size thirteen or fourteen – the foot size of basketball player Michael Jordan. In neighbouring Vindolanda, only 0.4 per cent were that size. This implies the particular ethnic regime in that fort would have towered over the local population.

17 January 2015

When I was eating in the Gate at Yanwath, I heard a lady at the next table remark loudly, 'Of course we are all Vikings here.' I looked up. And everyone at her table nodded, unsurprised. But two hundred years ago, Cumbrians would have been very surprised to learn they had Norse ancestors.

One of the first clues was picked up by the editor of the *Westmorland Gazette*, who in 1819 heard a woman imploring her naughty child, to 'Come its ways, then, and get its patten'. It was only because his brother had been imprisoned by the Danes for eighteen months that he recognised she was using the Danish word for 'breast'.

It was Victorian scholars who began the process of recording in detail just how many Scandinavian words survived in place-names: 'fell' (fjall), 'dale' (dalr), 'beck' (bekkr), 'gill' (gil) and the rest.* This became increasingly apparent from Cumbrians studying Icelandic, and from Norwegians visiting Coniston. Scholars began to look more carefully at eroded figures on Cumbrian crosses. And by the end of the nineteenth century, some traditional Cumbrian dialects began to be revealed as being closer to Old Norse than to modern English.

* Words familiar to all Cumbrian readers but perhaps not to all readers – meaning a high open moor, a valley, a small stream and a narrow ravine.

Finally, ten years ago, Penrith was revealed to have 'the highest concentration of Scandinavian DNA in England'. For more than a quarter of us, it seems, our times-forty male ancestor was from Norway.

And yet, this had never been how pre-Victorian historians had described our area, or how many modern historians continue to describe it. Roman, yes, part of the Anglo-Saxon kingdom of Northumbria, perhaps even partly Scottish or, for the pedantic, part of the Welsh-speaking kingdom of Cumbria. But the Norse were not recorded in written chronicles. Nothing in the historical record, and very little dug from the ground, suggested this had been a very Norse area.

And yet it was. Place-names, Cumbrian dialects and crosses confirm that some time at least a thousand years ago, Scandinavians settled almost every valley in the Lakes from Wasdale (Vatnsdalur or valley of the water) to Ullswater (Ulf's water). But we have to rely on much later sagas from Iceland to even sense their values and culture. It seems that the Norse who came here were no longer only 'Viking' pirates: although they may have continued seasonal raiding. They came here to settle: bringing their families, their livestock. They came to Cumbria with enough power to obliterate almost all the earlier place-names of the 'original inhabitants' and replace them with their own Norse names. They took over small upland farms, kept pigs, and focused on cattle and sheep. 'Herdwick' – as in Herdwick sheep – comes from the Norse word for sheep pasture. They coppiced ash trees to feed their livestock, and they built dry-stone walls.

Their society was probably more democratic than the medieval world that followed it, reaching decisions – like other Norse societies – through assemblies on mounds called 'thing-mounts'. But they also kept slaves (our local village of Threlkeld

means 'the place of the Viking slaves'). If they were like their
Icelandic cousins recorded in the sagas, they could be quick to
take offence, fiercely independent, obsessively litigious (suing
their neighbours over land boundaries, grazing rights and inheritance) and willing to kill each other if the disagreement went
too deep.

All this might make the Cumbrian Norse, 1,200 years ago,
seem 'backward' and parochial. But they had global connections, which few Cumbrians have today. Their fathers had sailed
from Scandinavia around the Orkneys. Their contemporaries
and relatives fought on the coast of Libya, served in modern
Istanbul, set up dynasties in Kyiv, and even established a small
settlement in North America.

Unlike the British or English locals, these new immigrants
were not Christian when they arrived. They took their names
from Norse mythology; from Odin and his sacred bird, the raven
(hence Oddendale, Crosby Ravensworth). They believed, it
seemed, that when they saw a Cumbrian rainbow stretching
over the fell, they were gazing at a bridge at whose end lay the
home of the gods. When they watched the sun cross the sky,
they believed its light was being dragged by demigods, galloping
in terror from wolves at their heels; and that the wolves would
eventually catch the sun and plunge the world into darkness.
They expected that their society, the world, and the gods themselves would all one day be wiped out in a great apocalypse.

Once we have become aware of this lost community, we
cannot unsee them. The stone on Kirkby Stephen churchyard
is a cross, but look more carefully and you can see that the god
Loki is carved, bound in the entrails of his children, beneath
it; their sacred ash tree (Yggdrasil) is on a sandstone pillar in
Dearham church, and their world serpent is carved on a stone
in Penrith churchyard.

These stones were probably used by Christian missionaries to convince the Norse that Christianity was not very different to their old beliefs. (Loki and the serpent were just versions of the Devil, perhaps, the beautiful and peaceful Baldr was Jesus, and the ash tree was the cross.) But today our Norse ancestors should remind us not of similarities, but of deep and startling differences. It is tempting to get round the strangeness of the past by making it simply glamorous. You could point out, for example, to the lady who said 'we are all Vikings', that her forty-one-times great-grandfather might seem familiar: a tall dalesman with an upland farm, a Herdwick flock behind his dry-stone wall; but the silver on his brooch came from Islamic coins minted in Samarkand, the bangles recorded the number of enemies he had killed, the hammer round his neck was an amulet of Thor. And it is easier to describe his jewellery than his mind; or to inhabit his values, his honour and sense of justice. We should hold back from claiming to understand. Why did such great care go into depicting Thor and the final apocalypse of the gods along the fourteen-foot shaft of the Gosforth cross? What did it mean to be an upland farmer who also believed that if we die in bed, instead of battle, we go to hell?

22 December 2012

Christmas is, among many things, a reminder of how Cumbria was the child of an ancient nation, scarred by the Romans, invigorated by the Norse, but developed by the Church. When the Romans departed, they left chaos behind. Penrith became part of a Celtic kingdom with borders hovering around the edge of Clifton and Brough. Chiefs, whose ancestors could read, chose to be illiterate; the Roman water systems of Carlisle were abandoned; the roads collapsed. In Birdoswald, a leader built an awkward wooden hall amidst the fine masonry of the old fort. According to the Cumbric epic poem, when an invasion happened, the Celtic warriors chose to feast and drink for a year before riding to be wiped out in battle near Catterick.

Then Cumbria became no man's land. Seaborne peoples arrived: Irish in Galloway, Anglians on the Northumbrian coast and, in the Lake District, the Norse. They have left traces in the Viking haaf-net fishing on the Solway; in the pollarded ash of Borrowdale, identical to those of the fjords; in the Norse ash tree and god Loki carved on the stone in Dearham church; and in our dialect.

But they lived in a poor, fragile scattering of hamlets. This was the time when miniature kings signed treaties at Eamont

Bridge, and Eric Bloodaxe died in battle on Stainmore. By the time of the Norman conquest, Cumbria hardly seemed to be part of any country. William the Conqueror, who wanted to document almost everything, did not record Cumbria in his tax documents. William Rufus of England built his castle at Carlisle; David of Scotland called himself Prince of the Cumbrians and held a parliament, but no nation had a monopoly of power in Cumbria: none perhaps wanted to.

It was the Church that saved us. Within a year of the death of a reforming monk from northern France in 1117, his followers had established their first community in the Borders. In 1123, another group set themselves up in 'the valley of the deadly nightshade' by Barrow-in-Furness. In 1138, Cistercians arrived in Melrose, and then established an outpost on an island, surrounded by a hundred square miles of wetland bog, beside modern Abbeytown. And in 1199, a fourth order made a church for themselves just below the bare high pass of Shap.

They took a wilderness and created a garden. The abbeys of Furness and Shap inherited mountain, bogland, heather and scrub. They felled trees, drained mires, cleared the bracken and gorse, and cropped the grass. They laid mile upon mile of dry-stone wall. They developed better breeds, and they hefted the sheep to the hills. In West Cumbria, the monks of Holm Cultram drained the marshes and created, out of a treeless expanse of heather, bog myrtle and sphagnum moss, fields for oats and barley, pasture for dairy cows, and dry land for majestic trees. The abbey of Furness developed mining; the monks of Holm Cultram developed a port.

From Melrose to Furness, and from Shap to Dryburgh, the churchmen created a new economic system and a new society. They developed an export market in wool, which turned

Berwick into one of the great trading stations of its age. They led a renaissance in crafts and architecture; and employed thousands through the extreme ambition of their building projects. They introduced the first standard taxation and land assessment systems. They provided care for the sick, schools for the illiterate, and shelter for travellers.

These men came to the Borders because they wanted to live outside society. But they introduced us to a global conversation. Their mother monasteries were on the continent. Their abbots included Italian, Austrian, Flemish and British men, who had studied in the great capitals of Europe and established foundations in Scandinavia. They included some of the greatest writers of their age, and they went on to support universities; but alongside their lay brothers, they worked all day through the summers clearing stone from our fields.

And they did so in rough cowls of unbleached wool, eating once a day in the winter, sleeping very little, removed from their relatives, forbidden privacy, or families, or personal possessions. They chose to live, work and die in the Cumbrian fells and the marshes.

Today, Barrow and Shap, Holm Cultram and Melrose, Dryburgh, Kelso and Jedburgh are ruins. In each place only slender stone facades remain. It is tempting to interpret this as a symbol of defeat. It can seem the beginning of a long trend of falling congregations, of unwelcome developments and innovations that undermine all that is sacred and good. But the abbeys and the history of the monks are a reminder not of the vulnerability of the Church, but of its power, energy and engagement in shaping the modern world.

We are the heirs of these childless men. They can be traced in the close-cropped fellside, and the limestone wall at Swindale, and the oaks by Abbeytown. What we understand of our ancient past comes from the learning they preserved. We should remember this Christmas that, over five hundred Christmases, in weather like this, these churchmen, from lauds at daybreak to vespers at dusk, laid the foundation of our Cumbrian civilisation.

10 May 2014

> . . . no English man, that has any honour in the glorious memory of the greatest and truest hero of all our kings of the English or Saxon race, can go to Carlisle, and not step aside to see the monument to King Edward I. at Burgh upon the Sands, a little way out of the city Carlisle, where that victorious prince dy'd . . . that prince being the terror of Scotland.

So writes Daniel Defoe, the author of *Robinson Crusoe*, in his guidebook of 1726.

You can still see the same sandstone pillar in the middle of the salt-marsh, and it is almost certainly built on the exact spot where the king died, aged sixty-eight, in 1307, having been given dysentery by the Cumbrian water. It stands very close both to Hadrian's Wall and the modern border. It is easy to think of the man who died there as an English king who had fought two Scottish kings; and to see the pillar, like Defoe, as the ultimate symbol of the division between England and Scotland.

But Edward's normal language with his friends was not English; it was Norman French. And that was true, too, of his rivals, the 'Scottish' kings, John Balliol and Robert the Bruce. Edward was not a Saxon and they were not Gaels. All their paternal ancestors were French nobles who had come to Britain within the last two hundred years. Jean de Bailleul meant he had come from Bailleul in Picardy. The second king, Robert de Brus, means Robert from Brus, a village on the Cotentin peninsula of Normandy. And Edward himself was, of course, the heir of the Duke of Normandy, William the Conqueror.

What we now call Scotland and England were then occupied by a medley of ethnicities: the descendants of British-Welsh, Northumbrian Angles, Norwegians, Flemings, Norman French, Danes, and more. The Scottish Declaration of Independence at Arbroath shows that the Scots believed themselves to be immigrants from the edge of the Black Sea and Spain, who had exterminated the previous populations who once lived north of the Solway. But most of the men who signed the declaration were descended from families that had come relatively recently from modern France (and in one case Hungary).

If you stand at the monument and look across the Solway, the low rising ground of Annandale in front is the land of Robert the Bruce. To the west in Galloway is the land of John de Balliol. But behind you on the 'English side', there is more of their land; these pretenders to the throne of Scotland had almost as many men and acres in modern England as in modern Scotland: great estates in Yorkshire, land around Penrith. The family of the 'Scottish' king John Balliol endowed the college where I studied in Oxford; Barnard Castle in County Durham takes its name from Barnard de Balliol. And they still owned the town of Bailleul in France – and indeed King John of Scotland was to spend the last dozen years of his life there.

Since both Bruces and Balliols held so much land between the realms of the English and Scottish kings, they seemed to be with both and neither. At a twelfth-century battle, both Bruces and Balliols fought against the Scottish king on the side of the English king, Edward's ancestor.

So this was as much a fight for power between Norman-French cousins, as a battle for Scottish independence. But the family nature of the feud did not reduce the horror of its consequences. When these cousins began to fight each other, the killing hardly ceased for the next three hundred years. Edward I massacred the people of Berwick, and hung Bruce's female supporters in cages from castle walls. Bruce's raids south cut down all the trees in orchards, sacked holy shrines, and destroyed the soil so that the population could not plant crops again. Bruce, whose family owned estates deep into Yorkshire, now banned cross-border landholdings. He crossed over the Solway to Holm Cultram abbey, in Abbeytown, which his family had long supported and in which his father was buried. But because it was now a few miles on the wrong side of his border, he ransacked his father's resting-place and burnt it to the ground.

But all these men, however, would rather have been fighting in the Middle East. Edward had taken Balliol's brother and Bruce's grandfather on crusade to Syria in 1270. There, near Damascus, Edward made his reputation by killing, with his bare hands, an assassin who had broken into his tent. Edward remained desperate, thirty years later, to return to crusade, as soon as he had finished fighting Bruce. He had raised a fortune for the expedition.

One chronicler says that his dying wish was that his heart should be carried to Jerusalem. That was Robert the Bruce's dying wish, too. He too had wanted to go on crusade like his grandfather. But he could not make the journey because he had leprosy. Instead, he asked his friend James Douglas to cut his heart out of his chest and carry it to Jerusalem. James died cut down by a Muslim army in Spain, having flung the heart into the midst of their cavalry.

Defoe tells us to see the pillar at Burgh by Sands as a symbol of the 'English or Saxon race' fighting the Scots. And of course, in the movie *Braveheart*, Edward is English and Bruce is Scottish. So, in the year of the referendum, it may be worth remembering that Edward and Bruce would not have been people whom we would recognise today as English or Scots. They were French-speaking noblemen, descended from Scandinavians, and related to each other. Much of their energy was directed towards Palestine. Edward, one chronicler says, had been warned that he would die in Burgh. But he had thought Burgh was a name for Jerusalem. He died enraged that it turned out to be a hamlet near Carlisle.

21 December 2013

Five years ago, walking around Cumbria, I was mesmerised by the echoes of the past. In a section of the Lowther valley, only one new house has emerged since 1700: the white farm-buildings, built into the steep slopes, evoked a remote community in the Indian Himalayas.

Nature was insistently present: in the ash leaves, slowly emerging; in the cold clarity of the night skies; and the sound of ewes in the morning. So was pre-history, in the 3,500-year-old stone monument nestled in an amphitheatre of tens of thousands of river stones, at Mayburgh.

It was not difficult to imagine the Jacobite army, whether chasing the Bishop of Carlisle along the Lowther ridge-line or retreating through Clifton. This landscape of isolated villages and uplands, deeply marked with history and conflict, stretched from Cumbria and Northumbria to the Scottish Borders. It was the landscape of a Middleland, an autonomous zone, not part of either England or Scotland. And its elements, Anglo-Saxon, Viking and Celt, its green and often cloudy hills, seemed foundation stones of a much broader British identity.

So I went on a longer walk through Cumbria and the Borders last summer, hoping to uncover at a slow walker's pace the liveliness and richness of British identity.

At first, the echoes of soil and landscape were, almost, what I had expected. Eric the farmer at Ullswater, Barry the Blencathra huntsman, and Ron, a retired horse-dealer in Dearham, were strongly shaped, it seemed, by the places in which they worked. I walked past the burned shell of a medieval abbey and then met the family whose son had burned it. I saw how the lands that monks had drained were being re-flooded. I forded the Solway with Mark, who fished in the Viking haaf-net tradition, and chatted to Duncan, who continued to celebrate the Border reiver tradition in everything, including his son's name, 'Reeve'.

But, on entering Scotland, such moments seemed more misleading. The Borders was one of the most remote, sparsely populated areas in Britain. But it felt startlingly divorced from history. The crofts and family farms were almost all long gone, replaced with large farms whose tenants had often moved to the area only since the Second World War.

A landscape once defined by medieval castles, monasteries and conflict was now hemmed in by the council-tended banks of the slow-flowing Tweed. The traditional figures of rural life – farmers, labourers, landowners, priests, doctors, schoolteachers, policemen – seemed to have shrunk, changed or vanished entirely. Most people were either commuting or had retired, so their memories or interests were in another place. More than 90 per cent of people had not been born in the village in which they lived. There was little oral history, and very little interest in, or affection for, British institutions – least of all Parliament. British identity seemed very weak. Although Northumbrians often told me they were honorary Scots, no Scot repaid the compliment.

People often felt powerless and marginalised, never quite able to bring their idealism or skills to use, stuck with jobs and governments and an environment they disliked, but felt unable to change. Few seemed to feel the deep, rich connections to landscape, history and institutions which once underpinned British identity.

There was profound loss of the past, of cultural structures, of traditions, values, of local particularity and, above all, of confidence. And, then turning south again, the same seemed true of many Cumbrian villages. It was disheartening.

Yet almost every door in Cumbria was also a revelation. You could not guess which countries people had visited, where they had lived, or what concerns they had (interests included American model trainsets, fair-trade in the Leeward Islands, and the design of apps for river-walkers). There were no Christian monks, but there were strong communities of British Buddhist monks from different Tibetan and Japanese traditions. People had never been so educated, healthy, or highly, unpredictably, individual. Everywhere was the pulsing presence of women and men whose lives, interests and tastes were almost entirely different to those of their parents.

To accommodate these new talents, energies and imaginations, we need a political framework capacious and plural enough to liberate tens of millions of individuals. We need a context which allows separate local identities to flourish alongside very different cultures: a context that would encompass Cumbria, as well as Aberdeen or London. A new idea perhaps of Britain.

26 April 2014

This land was first turned from bog into farmland in the twelfth century. The medieval monks cut drains, burned back reeds and choked the sphagnum moss to create rough pasture. By the nineteenth century, it had become a landscape of hawthorn hedges and oak avenues, cut with rich soil. Now, on the field edge, the whitened trunks of the oaks were drowning in three feet of standing water. My companion was 'turning the clock back'.

The Royal Society for the Protection of Birds (RSPB) began buying the farmland near Bowness-on-Solway in 1988. In Cumbria, it also acquired a cliff habitat for sea birds; an artificial reservoir for lake birds; and sections of high and low fell. Half a mile from us, on the estuary, were half a million wintering waders, 'bar-tailed godwits, black-tailed godwits, golden plovers, grey plover, ring plover, oystercatchers, dunlins, turnstones'.

When the RSPB bought this land, it was a dry heath, cracked into deep gullies, and filled with cattle. They began by blocking the old header-drains, and used the remaining pipes not to drain but to flood. The initial aim was to make sure that you could push a six-inch nail easily into the soil. 'That tells you how easy it would be for a wading bird to probe its beak in to find food.'

But different birds needed different wetness of ground. Snipe preferred a wet, rushy habitat with a few dry bits to lay their eggs on. Lapwings, however, wanted closely cropped sward and drier conditions. 'But they do like the odd tussock dotted about, that they can get tucked behind, to hide. And small pools with lots of muddy edge so the chicks can get to it and feed.' Redshank tended to prefer it 'more tussocky with more clumps of rush: but not wall-to-wall rush'.

Getting the right length of sward for the lapwings required the RSPB keeping its own flock of sheep, contracting with nine different graziers, and experimenting with native-breed cattle. Every August the grazing was supplemented by machine-cutting and an artificial raising of the water-level. 'If you can get the water just above the cut stems of rush, it tends to kill them off, but it's an art in itself timing it just right.' Curlew wanted intact mires. He took me out into the bog. 'What we're standing on here is something like 95 per cent water. Milk's got more solids than this.'

Bogs don't have a good name – our Northumbrian ancestors put monsters like Grendel in such places. The border reiving clans used the bogs or mosses as a hiding place and a trap for their enemies. They called themselves 'moss troopers'. For centuries, they had cut long paths and ditches into the wet ground to extract peat for fuel, and then allowed the vegetation to hide the paths and twenty-foot-deep ditches.

In the centre of the mosses lay deep ponds covered in small rafts of floating sphagnum. Unwary pursuers could be lost in the bog. 'We're all right while we're just treading carefully but I've had animals like cattle: they've got out onto the bog about thirty yards, and they're just thrown. They start sinking up to their bellies. There was one in particular, it just wouldn't move. We had to drag it.'

But my companion saw the bog as a paradise. 'About 94 per

cent of our lowland raised mires have disappeared, so we're just working with the last 6 per cent.' He showed me the bog asphodel, the stag's horn lichen and the white beak-sedge, 'one of the foods of the large heath butterfly'. He pointed to the round-leaved sundew: 'Because this habitat's so nutrient-poor, it has to supplement its diet by catching flies. So these little sticky tendrils on the leaves: the fly lands on that and it catches the fly and slowly absorbs the nutrients from it.'

His favourite plant was the bog-rosemary. 'When I first came here and started monitoring the site I found five plants of bog-rosemary in a whole day of searching. Now we've just mapped its distribution right across the bog. It's just so beautiful. If I'm doing surveys at night when everything's out, I just crush that up and rub it over my face. Just break it up and crush it,' he encouraged me, 'have a whiff. It keeps the insects off to a certain degree. I wouldn't say that it's as good as the juice you get in a can but it's the next best thing.'

Now he wanted to extend the bogs. 'Given time, when we've got the nutrient levels down, tried to get rid of some of the phosphates, the sphagnum moss will come back, which enables the peat to grow, and then that'll eat up more of the nutrients, and you'll eventually get bog flora starting to come through.' This would provide even more space for the curlews.

'Amazing. And how long do you think that will take?'

'I'm not going to see it.'

'You're not going to see it?'

'No. Nobody really knows. I mean, a lot of what we're doing here is quite groundbreaking . . . there's been some remarkable changes in that fifteen years but nothing suggesting we're getting much closer to creating ideal conditions for bog flora. So, fifty years maybe?'

Decades of flooding, mowing, grazing and monitoring

has transformed the 800-year-old agricultural landscape and produced sixty-five pairs of wading birds across the thousand-acre site.

'When we moved here twenty-two years ago we didn't have any. Zero to sixty-five over the years,' he said with pride. 'Everywhere else in the country is going down.'

London, 14 May 2011

This 'interlude', like the four others that come later, was also a letter originally published in the Herald. *They are called 'interludes' because unlike the other letters they do not fit into the chapters on Cumbria.*

The centre of London on the day of the wedding was cut by police barriers; the Underground stations were closed; it felt hot; and there was someone on every paving stone between Millbank and Leicester Square. It took me an hour and a half to find a route through the crowd back to my aunt's flat, and I just had time to drop my tail-coat and pick up a toothbrush before heading back to Euston.

Once home, I was asked about the wedding in Pooley Bridge, Crosby Ravensworth, Appleby, Tebay and Kirkandrews-on-Esk; in Longtown, Temple Sowerby and Warwick Bridge. I didn't know what to say.

I was quizzed intensely, for example, at a royal wedding street party in Shap. The grandparents were wrapped against the cold wind, the parents were enjoying the bright sun, the children were riding round and round on bicycles, and on the edge of the crescent of houses, lambs were grazing. I was given a slice of Union Jack cake, and someone politely asked the questions which my mother asks about every wedding.

'What did you think of the bride's dress? Who did you meet?' But they were far better informed than I. They had seen it through cameras flying down the buttresses and zooming in on the princess's ring-finger, with an expert commentator who could tell the Agong of Malaysia from the Sultan of Brunei. Everyone agreed that Catherine's mother's dress was perfect; that Princess Beatrice's outfit was not; that Harry was fun; and that the couple were very much in love.

But I hadn't noticed any of those things. I was seated two rows back in the nave, behind the broad scarlet coat of a Yeoman of the Guard and a tree: a slender field maple with bright green leaves, with lily of the valley at its base. Oh, and I saw beech in the transept. I was used to weddings where guests pour in at the last moment, ushers try to seat giggling long-lost cousins, old men laugh over organ music, mothers try to soothe babies, fathers carry them outside, and where there are children, in white, under everyone's feet.

This was not that kind of wedding. We arrived more than two hours before the service, each of us with our own colour-coded ticket. I was seated next to the prince's farm manager and his wife. Our view was of a thousand wooden chairs, a ceiling a hundred feet high and monuments to forgotten heroes on the walls. It was cool and very quiet and there were no celebrities in sight. Facing me was a soldier in uniform, with many medals. He served, I think, with Prince Harry in Afghanistan. Behind him was a deer-stalker from Scotland. The old head of the prince's security team, whom I had last seen in Kabul, was wearing an immaculate tail-coat. We talked a little about dairy farming in Gloucestershire.

Someone who had known Prince William as a boy said she could not quite believe that he was old enough to be married: and she cried a little. But generally we sat in silence. There were no children and no flirting cousins, and I don't remember any organ music. Meanwhile, a heavily built African man in a velvet suit with a silver cane was led to his seat; the clergymen and choir-masters stopped pacing in the aisle; and the grand old men of the Queen's bodyguard, in their silver helmets and ostrich-feather plumes, shuffled into position.

When the court trumpeters played their fanfare, and the Yeomen of the Guard entered with their cockades in their hats, it seemed for a moment like the prelude to a comic opera. But that was the end of pomp and circumstance. Despite all the generals and uniforms, there were no marching men, clanking spurs, forest of swords, nor bugle calls. Despite all the inheritance of a 1,400-year-old crown, there were no rituals of Edward the Confessor, and no great reeling off of titles. Despite the copes and mitres, crosiers and choirs, the churchmen seemed formal but not priggish: dignified, without pomposity.

The Bishop of London's sermon was a subtle and testing reflection on love. And although every movement had been rehearsed and perfected in some cases by generations long dead, it was not lifeless. It moved like a solemn dance.

And now I'm back in my kitchen, exactly two weeks later. The blackbirds fell silent three hours ago and all I can hear is the mosquito-like buzz of the strip-light above my head. It's past midnight and I haven't said anything about the other days of the fortnight: the local elections, or the anniversary of my first year as an MP, or my father's eighty-ninth birthday.

Yesterday, I brought the Skills Funding Agency and Askham Bryan College together to talk about Newton Rigg, and we made some progress. The willow warblers are back from Africa and there are whorls, like miniature beehives, on the walnut twigs. Scotland may vote for independence. I am introducing my first proper motion on the floor of the House next week, which will, I hope, bring mobile coverage to 2 million more people in rural Britain. And bin Laden has been killed. But Joan Raine, the chair of Crosby Ravensworth parish council, has just asked me to speak about the wedding.

STATE LAND

T HESE ESSAYS ARE PARTLY ABOUT IGNORANCE. I first encountered our staggering lack of knowledge in Iraq and Afghanistan, but it was as a Member of Parliament that I discovered what a scandalous and inevitable feature ignorance was in a minister's life. And the many ways in which civil servants struggled to fill that gap.

Government statistics made Penrith and the Border's economy almost unintelligible. Average farm incomes were, as I have said, £8,000. Ninety-two per cent worked for businesses with fewer than ten employees. Behind these numbers lay a dense web of micro-enterprises that defied every model of economic development. Walking down a village street, I might encounter a journalist, the organiser of a beer festival, an environmental consultant, the manager of a holiday let, a wellness coach and a retired schoolteacher from Manchester. Once I met a woman who was all those things at the same time. The local economy seemed defined less by management, efficiency and technology and more by relationships, local knowledge and extreme adaptability. It was strongly shaped by geology, topography and soil (which had kept coal outside my constituency and upland sheep pasture within) and it was distorted by many misconceived government industrial policies.

The majority of my constituents voted Conservative. But, as these letters show, this had little to do with the Thatcher–Reagan slogans of privatisation, deregulation and globalisation.

These hard-working entrepreneurs in small farms and agricultural businesses were mostly kept afloat through high tariffs on agricultural imports, and tens of millions of pounds of EU subsidies for single farm payments and environmental schemes. The overwhelming presence of government regulations and subsidies did not, despite the predictions of my colleague Liz Truss, create inertia and dependency. Instead, people worked very hard in businesses – like upland farming – which central statisticians would have perceived as inefficient and unproductive. And they volunteered to provide many services which in other areas government was expected to deliver.

These essays record my attempts to describe how the economy of the north-west worked: why policy-makers often couldn't see what was happening on the ground, and what happened when London tried to help. It begins, however, with another walk, some glimpses of politics in the centre, and some history.

1 September 2012

Yesterday was the end of five weeks' walking in Cumbria and the Borders. It was a thirty-nine-mile day, and twelve of those fourteen hours were spent in almost empty space. In a car I can be transported at a mile a minute from centre to centre, from one room with people, a timetable and a purpose, to another twenty miles away: from the George in Penrith to Appleby Grammar School, from the munitions depot at Longtown to the Local Links centre in Wigton. But walking makes each yard of ground equal, draws you into the space between centres – which is now often unpeopled.

It once seemed as though industrialisation would replace the countryside, littering it with brick walls, concrete paths and people. In the 1930s George Orwell gazed horrified at the rotting detritus, the scars and the smoke spreading across the north. But walking reveals that the last two hundred years had not filled so much as emptied a great deal of the land.

So, on this journey, my walking companions – farmers, and officers from Natural England and the Environment Agency; archaeologists, nuclear activists and painters; schoolteachers and doctors; climbers and pensioners – joined me at villages. Between villages, I saw almost no one. Walking alone, I could note the swathes of Yorkshire fog and cock's-foot grass, or the

thick cropped form of an oak stubbornly wedged on an abandoned dyke. I could wonder whether this now marshy field had
once been drained by the Romans, or by the monks of Abbeytown, by Mr Curwen the Georgian improver, or by post-war
subsidies; and question why the ragwort was flourishing or why
the bracken was turning early. But there was rarely anyone to
answer.

It would have been different not so long ago. The now
tranquil, silent country by Haweswater or above Glenridding,
by Caldbeck or opposite Threlkeld, was then filled with the
hammers and detonations of miners and quarry-men. When my
father worked in the Forestry Commission during his summer
holidays, a hundred men, with axes and saws, did the work now
performed by two harvesting and forwarding machines.

The village of Bampton, which now lacks any business apart
from the pub, included, when my grandfather was in his thirties,
cobblers and carpenters, weavers and many small shopkeep-
ers. There were six farms for each farm that survives today, and
small farms typically employed four people. So perhaps twenty
times the number of people were farmers. In Swindale there
were enough houses for a church, where there are now only
two families and the faint outlines of ruined cottages. The tracks
between valleys, now deep in heather, would once have been
pounded by people carrying, visiting, running errands. And you
would have seen and been greeted by dozens working in the
fields and along the fellside.

It is still like that in some parts of Asia. A remote boulder, at
4,000 feet, will act as a throne for a shepherd watching a dozen
goats grazing the bare and flinty soil. When I was walking
through Iran, I found that almost every orchard seemed to
include a man on a carpet with a battered metal teapot. In Nepal,
men in peaked hats and tight cotton trousers often sprinted past

me on narrow tracks. I came across tea-houses built on stilts for walking villagers, five days from the nearest road. Countrypeople could reel off the qualities of streams, the positions of trees, the names of men twenty miles away. In the winter, I watched farmers steering the oxen and the heavy plough through the mud, and in the autumn saw the animals circling slowly around the stone threshing floor, and the golden cloud of grain flung into the air from winnowing trays.

In Cumbria now, even in August, many of the great national trails are unused. It sometimes felt as though the millions of pounds of trails and signs and stiles were used by nobody except me. All the arcs of the ridge-lines, the ash branches alive in the wind, the flocks and the herds were unseen by human eyes. It is no longer worth someone sitting all day to guard a hundred sheep. More money can be made from selling a farmhouse as a rural mansion than from agriculture. Government policy and subsidies and schemes have paid farmers to stock less, use less ground, turn more and more over to wind turbines, or to nature.

We have of course achieved miracles in undoing the damage of industrialisation: the meadows have re-emerged over landfill sites and open-cast mines; and native trees have been planted along the riverbanks and motorways. But the loss of the human presence is dramatic and troubling.

This is a land where centuries of effort, of stories, of habitation can be traced in the shape of soil. It is the farmers – and the lambs they keep, and the barns they build and use, and the fields they manage – which still just make Cumbria something more than what so much of Scotland has become: open hills of overgrown heather beneath turbines, blocks of commercial forestry, and occasional glimpses of pale pink patches of sphagnum moss on re-flooded peat.

4 July 2015

My new ministerial office is a great grey stone building, in a square near Parliament. You enter beneath some scaffolding, under the sign 'Department of Environment, Food and Rural Affairs', walk through the crowds by the lift shafts, up six flights of stairs and emerge onto an oak-lined corridor.

The corridor is empty. My office is packed with never less than five people, but often fifteen, either explaining a decision or offering a choice, often many choices in a day, which we then have to present to the Secretary of State.

Neil came in today to discuss funding local flood schemes, and insurance; how to deploy 5,000 extra staff in twenty-four hours for a flood emergency; the state of the sea-walls on the Humber; Grimsby's role in offshore wind; and flooding forecasts. He explained that the Met Office and its super-computers were getting better all the time: last year they predicted a coastal surge early enough to evacuate, so nobody was killed, whereas an almost identical surge in the same place in 1953 killed hundreds. Neil explained that the most likely problem for us this week would not be coastal flooding, river flooding or ground-water flooding, but surface-water flooding.

'Where?'

'We can't tell yet.'

'Summer thunderstorms,' added the man from the Met Office, 'are like boiling a pot of water – you know there will be bubbles but you can't predict where.'

Next up is Tom, a tall, tanned, smiling man who has managed to persuade the department to let him work remotely from Bristol. He has travelled up on the train to ask some technical questions on parliamentary legislation and give an update on the national parks. On the way out, he summarised the 160 different types of landscape character which have been identified in England; told me about cricket in Bristol on a weekday afternoon; and explained why Arthur Ransome loved the Norfolk Broads.

At the other end of the room, Jo was reorganising the seventh meeting of the day, working out how I would get from negotiations with the Royal Society for the Protection of Birds, in my office, to a meeting about village halls in the Cabinet Office, twenty minutes' walk up Whitehall. Liz was making some suggestions to Chetal for a speech on recycling; and Suzie was preparing for the European Union negotiation, which involved making sure that I would be in the right corridor in Strasbourg to catch the attention of the German State Secretary, and query the commission's proposals on the habitats directive.

Cheryl, meanwhile, was telling me what atmospheric emissions ceilings I should agree to set for 2030. Questioning Cheryl forced me to try to understand exactly how ammonia concentrations could interact with PM2.5 in the air, and the potential impact of air-quality ceilings on the grass-fed dairy industry. Ben was sitting beside me, trying to work out how we could communicate this science to colleagues and voters.

Tom, who was tired from replying to my emails at midnight, was now dealing with the fact that 20,000 people in Hampshire had just lost their drinking water. He said that a vital spare part had been flown in by helicopter, but apparently the part hadn't fixed the pipe. Plans were being put in place to distribute emergency water to the elderly.

On my desk is a box of sixty rowan, cherry and birch saplings – a sample of the millions which we are proposing to distribute for schoolchildren to plant. And Tim, who procured them, is elbow-deep in the box. When I have finished on air quality, Tim explains the many ways in which the children could kill the trees, and particularly what could go wrong with unsuitable soils, poor staking, or too much access for deer and rabbits. Then it is Geoff's turn to explain the new model which he has helped to develop on Natural Capital. The model tries to calculate the value of a forest, for example, not simply for the commercial price of its timber, but for the carbon that it captures, for the wildlife it supports and the joy it brings to visitors. Two immediate problems: how do you calculate the number for joy? And even if you could, how do you get the Treasury or someone to pay for things which cannot be sold?

Almost everything in this wavering, staccato conversation with me – the new poorly informed minister – has a direct impact on Cumbria. In fact, no department concerns this place more than the Department of Environment, Food and Rural Affairs. And I haven't got beyond the Environment part. My predecessor's decision about the 'biodiversity 2020 targets', for example, explains everything from the location of a surreal fence line in Mallerstang to the closing of the peat works near Bewcastle. Our decision on flood insurance will determine whether it is possible for people near Eamont Bridge to get a mortgage. And the two thousand people in this constituency employed in the wood industry will be profoundly affected by my decisions on the ownership and value of forests.

At night, I have been reading *Thinking, Fast and Slow*, a book by Daniel Kahneman on human brains and decision-making. He shows from experiments how much we are affected by tiredness or stress, and influenced by half-forgotten memories; how difficult it is to make reliable calculations on risk, or on very long-term events. Which brings me back to the issues those fifteen people are struggling with in my office.

Even before one thinks about the context – the politics, the pressure groups, the media stories, the public misunderstandings which can hurt even the best policy – the complexity is staggering. I am being asked to make decisions on bacteria biodiversity, and on new scientific evidence (such as the precise impact of nitrogen dioxide on health); to sign off on climate models, long-term projections on technology and air emissions, and assessments of one-in-a-hundred-year flood risks; and make billion-pound choices on flood defences and water treatment. My decisions on 'resource management' and waste alone will help shape an industry employing 100,000 people. And all this has to be weighed against those things which are difficult to assign a market price to: our landscape, our traditional rural culture, even our sense of beauty and tranquillity. How – bearing Kahneman in mind – would a minister create the time, the challenge and the discipline, to think effectively about so many things?

29 March 2012

The truth, of course, is that politicians don't know what we're talking about. I don't mean that we are all stupid, or lazy (although I can sense my father's arched eyebrows, as I make that claim). But I mean that it is impossible for politicians to know enough. The most successful of us are, of course, brilliant at concealing this: we do our best to assimilate quantities of data, remember impressive statistics, and sound convincing on debt and drought, on customs and crime, on Inner Asia and inner-ear disease. But watch us carefully at the despatch box, or on *Question Time*, replying confidently to a hundred unexpected queries, never admitting that we don't know, and you must realise that not only are we bluffing, but that the whole system of departments, and parties and voters, and parliamentary questions and media interviews sets us up to be bluffers.

This is a problem. Because our best hope of making good decisions, or at least avoiding catastrophe, is for ministers to have the ability, confidence and, above all, knowledge to challenge bad policies. Our track record is not good. The US banks continued to invest in credit-default swaps for years; the European finance ministers let Greece drift for decades. We invaded Iraq, and sent more and more troops into Afghanistan. Again and again, politicians failed to realise that despite the confident

advice, optimistic predictions and encouraging figures, everything was going very wrong.

Take Tony Blair on Iraq. He was bright, his speeches showed
that he was well-briefed: full of obscure and precise statistics, confident about international law, clear about the global
order. But everything he had learned, everything he believed,
everything he expected about Iraq, was wrong. There were no
weapons of mass destruction, nor the terrorist links which he
imagined. He underestimated the strength and nature of the
Iraqi opposition, he missed the signs of civil war. International
credibility was not – as he predicted – helped by invading Iraq:
it was destroyed.

Now, of course, some people said so at the time. And some
were right for the right reasons, because they had an instinct
for the temptations of power and the errors of politicians, and
a few were right because they understood the detailed evolving
dynamics of rural Iraq. But it was easy for Blair to find apparently equally well-informed, thoughtful people who backed his
view: to find any number of sophisticated arguments, and statistics, which suggested he was right. Retired ambassadors protested, but none of his senior serving ambassadors or generals
formally challenged his decision in writing. Not one resigned.
Instead, they set about justifying and implementing his policy.
And he did not begin to have enough knowledge of the field to
even sense whose advice was more likely to be right.

The same errors, the same lack of challenge, exist in almost
every democratic system, and in the autocracies too. Look at
Germany's disastrous €100 billion investment in solar energy
(98 cents in every euro was wasted, and the effect, according to
the standard model, would delay global warming by twenty-
three minutes). Look at why steps were not taken earlier to
protect Cumbria from bovine TB; or the £38 billion debt run

up in the Ministry of Defence; or some of the more catastrophic investment decisions of the North West Development Agency. Ultimately all this was the politicians' fault; but in almost every case they were taking expert advice. They often followed the conventional wisdom of civil servants, which they lacked the experience, confidence or knowledge to challenge.

The solution to these problems cannot be to plug a bigger external hard drive into a minister's head. The sheer scale and complexity of modern government – from the NHS (with its million employees and 350 million patient encounters) to the procurement decisions on aircraft carriers – is ten thousand times beyond the capacity of any single mind. We cannot avoid, therefore, relying very heavily on the civil service. Every area – agriculture, small-business support, military procurement – relies on having officials who are well-informed, experienced and imaginative. Some need to be sector experts, others need the originality to ask questions which no normal person would ever ask, nor perhaps could ever answer. An official in the department of Environment, Food and Rural Affairs needs to ask what the thirty-year impact on farming would be of improving 'sites of special scientific interest', and what would happen if subsidies ceased to be paid. They need a grasp of Spanish animal movement databases, and an even clearer instinct of why we failed to predict the change in New Zealand's powdered milk exports. They must try to assess the possible results that might stem from a new technology that allows us to grow – currently at incredible expense – an edible hamburger in a laboratory from a stem cell.

To spot current follies, they need to have experienced the dangerous, unstoppable momentum of other fashionable theories in the past. They need to understand economics, and what economists don't understand. They need to be left in a department long enough to develop expertise, rather than being reshuffled between radically different departments every couple of years. And they need the courage to challenge not only politicians but also fellow Treasury officials, and many wealthy businesspeople with their powerful PR machines.

Perhaps, therefore, the best contribution we can make to the future of British government is to support the right kind of civil service. This is not a project that fits an election cycle: we need to recruit people today who may still be in the civil service in forty years' time; we need to give young people the incentive to get out on the ground, the time to think, and the nimbleness and courage to challenge conventional wisdom. We need the promotion criteria to ensure that those with the best policy judgement reach the top. We need to recognise when knowledgeable people have become rigid, or lost their desire or energy to fight. And we need to ensure that there is a culture of self-questioning.

Politicians are not encyclopedias, super-power memories, or saints. But we don't need to be, if we can build the right kind of civil service – and train politicians in how to listen to civil servants. But that is not enough. Because to do our job as politicians well, we also need to be able to sense when the whole establishment is digging itself into a hole, and while remaining aware of just how little we know, make that lonely decision to change course.

6 January 2018

The industrial revolution in West Cumbria was a miracle. In the second half of the nineteenth century, the slow growth of our traditional rural economy, celebrated by Wordsworth, was blown apart as we tore into the land, extracting slate and iron ore, and above all coal. In the fifty years leading up to the First World War, Cumbrian coal production quadrupled; pits, and then factories, provided employment for tens of thousands.

Cumbrian farmers poured from the upland farms of this constituency into the west coast, ribbons of brick nineteenth-century terraces emerged on bare moorland, grand munici-pal buildings dominated new towns – and the owners, at least, became very wealthy. While our part of Cumbria felt increas-ingly backward and isolated.

But the new industry of West Cumbria was far more fragile than the industry of the Midlands or indeed the American Midwest. It was already in trouble before German and US manufacturing began to outcompete Britain in the 1890s, and long before coal production and employment began its decline in 1914. Cumbrian factories were too far from major population centres. They were neither able to access the required quantity of labour, nor close enough to a mass customer base. We tend to think of deindustrialisation and rust belts as a story of

globalisation and China in the 2000s, or at least of Margaret
Thatcher in the 1980s. But this was a much earlier story. Most
of the population of West Cumbria was unemployed by 1930. It
was already clear that the county would struggle to sustain any
industries into the future.

A politician called Jack Adams attempted to resist this flow
of economic history. First as a Labour councillor and then as the
chief of the Cumberland Development Council and the West
Cumberland Industrial Development Co. Ltd, he brought government loans and guarantees into Cumbria for forty years. His
great energy and often a great deal of government money persuaded businesspeople such as the Hungarians Miki Sekers and
Tomi de Gara in 1938, and the Austrian Frank Schon in 1940,
to locate their factories in Cumbria. Adams helped reintroduce
industries ranging from silk-weaving to soap-making, and to
make Cumbria a base for the manufacture of fissile material for
nuclear bombs.

But the fundamental barrier of Cumbria's remote location
remained, and most of these enterprises ultimately collapsed.
In some cases, it was a change in global markets, in others, the
owners left when the subsidies dried up, and in many cases
the factory equipment was ultimately sold to places like India.
Some of the attempts to bring industry back were tragic. The
Williams coal mine at Whitehaven – famous for its bad venti-
lation and poor working methods – had become unprofitable
by the 1920s, and went bankrupt in 1933, by which time it had,
according to the chief inspector, 'probably the blackest record
in the annals of coal mining', with four major explosions in the
previous twenty years (in one of which 136 people had been
killed). Undeterred, Jack Adams, whose own father had been
killed in a mining accident when he was four, persuaded the
Nuffield Trust charity to issue a substantial loan to reopen the

pit in 1937. The new owners embarked on a highly expensive programme of improvements, at the end of which there was another explosion at the mine, killing 104 men. The National Coal Board closed the mine. Adams – in recognition of his attempts to sustain such industries – was made a hereditary peer.

The valiant efforts of Jack Adams and his many successors – politicians, development officers, treasury officials – have not succeeded in creating enduring industrial prosperity in West Cumbria, despite a century of effort. It remains one of the most deprived areas in Britain. And only the special government cases of the nuclear and defence industries – still backed by billions of pounds from taxpayers – have been able in the long term to escape the pressures of the global markets.

Meanwhile, this constituency in East Cumbria, which received very little attention and almost no industrial subsidy, did much better than West Cumbria. There was no difference in the population – in fact, the people in the factories were often the cousins of the people who stayed on the fells. But our very lack of factories – once a source of our poverty – proved increasingly to be the secret of our success, because it preserved an unspoilt landscape, which attracted talented people to work and millions of tourists to visit. Our growth and employment did not come from large employers, but from tens of thousands of micro-businesses, which remain to this day largely unstudied and unrecorded.

The sheer diversity of these small businesses – from guitar design to drone mapping, from UV water purification to chocolate pudding manufacture – is the key to our area's resilience. The constituency is not dominated by factories employing many thousands of people, and is, therefore, much less affected by the collapse of single businesses. While West Cumbria has one of the highest rates of unemployment in the country, our constituency in East Cumbria has one of the lowest. And our life expectancy is now on average seven years higher.

But when planners talk about growth in Cumbria, they still talk in terms of large factories, as though Jack Adams had never died, rather than understanding the economy we really have – which is based on an upland landscape, small farms, high-quality food, heritage, micro-business and high-end tourism.

5 July 2014

In all our market towns, councils are trying to create more growth. They have built more industrial parks; and then, claiming a labour shortage, more homes; then more businesses to employ the people in the new homes; then more homes, and so on and so forth. Historic town centres have become encircled by new housing and industrial estates.

The new superstores on the edge of towns have drawn people away from the increasingly isolated town-centre shops. There are more traffic jams and more pressure on services. The economy and the revenue for the council grow, but only because the population grows. The 'per capita' wealth does not increase; the individual does not become any better off.

Only three years ago, a very senior county councillor told me that he still dreamed of creating a car factory in the Eden valley. The car factory fortunately remains a fantasy. But his attitude was typical of almost all industrial policy in the north of England. This is not just true of the 1940s or 60s; we were still repeating the same mistakes in the 2000s. Almost £300 million was spent by the North West Development Agency in Cumbria. Much of it favoured large shops, large businesses and ever larger farms. Much was spent on grandiose multi-million-pound projects, many of which have left little trace behind. Confident arguments and theories continue to be recited to justify such strategies. 'We need to create affordable housing.' 'We need to create employment.' 'Surely you're not against growth?' These arguments are used even when the development includes hardly any affordable housing; even when there is very little unemployment in the area; even when communities are willing to build their own affordable housing; and even when the 'growth' does not increase individual incomes.

Insofar as Penrith and the Border survives as an area of diverse business, high employment, lively market towns and a beautiful, farmed landscape, it often seems to be despite, rather than because of government industrial policies. A patchwork of energetic, resilient communities, working in small groups, have often fought back against development officers and economic planners. Alston took the lead in preserving its historic townscape; Appleby resisted an out-of-town supermarket; Brampton preserved the GP surgery and care homes in the centre of town. The Friends of the Settle–Carlisle Line saved the railway from 'government efficiency cuts'. Upper Eden took control of its own planning policy through a public referendum. Crosby Ravensworth proved that a community could deliver high-quality, attractive, affordable housing with good heating systems and elegant design, without having to employ a big housing developer. The fact that our area still remains so beautiful, prosperous and distinctive is in large part because of community resistance.

But how much better we would do if the government, instead of imposing alien models, in fact tried to support the kinds of life, environment and activity which we want. Governments in other places have shown it can be done. France, Norway and Japan have all been much better at protecting a network of small family farms, local produce and local markets.

Italy and Spain demonstrate how historic market towns can be protected and sensitively developed. Even under huge pressure, poorer countries than Britain (whom we might expect to be vulnerable to development mafias and corruption) have banned out-of-town supermarkets; forced McDonald's and Kentucky Fried Chicken to use discreet signs and traditional buildings. Austria has been much better at bringing prosperity to small farms, preserving traditional architectural styles in the Alps, and encouraging young farmers to branch out into becoming guides for outdoor activities. In the Lake District, farming and our £100 million outdoor industry are kept very separate; in Austria such industries reinforce each other. Other nations have also managed to create a far higher-quality tourism offering, and attracted conferences and visitors in the 'off-season'.

We must therefore not be bullied into believing that there is no alternative to 'scale' and 'growth'. Instead, we should patiently explore and explain what we value about our area today, what we want to preserve, and what we would like it to look like in twenty years' time. We should not be afraid to say that some things are better left alone. Penrith and the Border is a place that thrives on beauty, community, small farms and small businesses. We need policies designed for who we are, not for what other people would like us to be.

11 October 2014

'I have no idea,' said my father, 'how Britain survives when we don't make anything anymore.' It's not a daft observation. When my father was twenty-five, British industry accounted for 41 per cent of the British economy. By 2013, it was just 14 per cent. Some 79 per cent of our national economy is now 'services'. What is Britain actually doing?

Producing food and manufacturing goods is relatively easy to understand. You can watch a farmer drain, lime, plough and plant a wild fell, creating food from wasteland. You can understand how technology brings agricultural and industrial revolutions. You could have seen a new plough, which could be dragged by two plough horses, replace one that required two horses and two oxen. You could have watched a mill take raw cotton and turn it into cloth that was sold in fifty countries.

But today's economy is not quite so easy to read. I spent last week in the Gilwilly industrial estate, just north of Penrith. At first, it could seem depressing – a 1946 government scheme allocated a patch of ground, adjoining some railway workers' cottages, for light industry. The earliest structures were little more than asbestos sheds on wasteland. Even today, you can drive in circles completely disoriented between the wire compounds and culs-de-sac. But Gilwilly, in fact, remains an instructive miracle.

By the 1990s, there were sixty businesses on the estate. Some still responded to traditional Cumbrian rural life, from animal feed to agricultural machinery, stables, wrought-iron fencing and equestrian supplies. Some to Cumbria's strengths in food. Other businesses were shaped by serving the long distances of the M6 corridor: haulage and warehousing companies, and the many businesses connected to vehicles (new cars, used cars, hire cars, 4x4s and Japanese superbike showrooms, new tyres and tyre repairs, car-washing, and spare parts).

Fifteen years later many of these M6 transport companies still survive, but the diversity around them is extraordinary – from the global to the insistently local. Stephen Armistead had spotted that containers come full from China and return empty, so he could ship used photocopiers as cheaply to China as he could to the south of England. He now has an award-winning green business, saving machines from landfill sites. Nearby, at Rebike, two unpaid volunteers are dedicating their days to training the long-term unemployed in restoring bikes. They are funny, straightforward and modest. The man finances himself with an engineering consultancy, and the woman by running a scallop factory in Annan. 'We thought we might make enough to pay ourselves, but it didn't quite work out, so we're still running it as volunteers four years later.'

Fylde Guitars, again hidden in an anonymous cul-de-sac on the edge of the estate, is, on the outside, a water-stained grey shed; on the inside, a treasure house. Gilwilly has allowed the owner to custom-build a £100,000 facility for gently drying rare tropical hardwoods. The shed houses brilliant pink snake-skin wood from South Africa, oily black planks of mahogany, pale 500-year-old Italian spruce, deep-veined rosewoods from Mauritius and Brazil, and logs of ancient redwood, dredged from Oregon rivers. He and his two craftsmen are working in techniques that would have been familiar to Stradivarius three centuries ago, slowly creating, over three months, hand-made guitars.

That is one kind of high-end manufacturing; JELD-WEN at the other end of the estate is another. Their production line in Gilwilly was custom-built for them in Spain at a cost of £5 million. Many of their 130 employees have been with them for decades. But they have retrained them from men working with hammers and nails, into computer and machine operators. This group of 130 people now produces more than a million doors a year – a quarter of the doors in the United Kingdom – and they are now beginning European exports. But delve further into this operation – with all its local Cumbrian energy and heritage – and you see it is in fact part of a US company, one of a hundred separate divisions of a firm which employs 20,000 people on four continents.

Keep walking down the estate and you will find another employer that is large by local standards: Greggs, employing 130 people to make gingerbread men and Christmas cakes. But the vast majority of the businesses here, and 92 per cent across the constituency, employ fewer than ten – a quarter are one-person bands. On and around the estate, I also see a voluntary FM radio station, a microbrewery, a canoe instructor, a sheep-wool insulator, a chocolate pudding maker, international governance consultants, renewable energy advisers and dance trainers for cruise ships.

I don't understand how they all survive. Something, however, is working.

Greggs and JELD-WEN are expanding their plants in Penrith because they find Cumbrian workers are more productive, and are producing at a consistently higher quality than their workers elsewhere. Eden farmers have won the awards this week for the best poultry farmer and the best dairy farmer in Britain.

One of the best things that a government can do for an economy so diverse and so bewildering is not to select individual factories to subsidise, but instead to provide shared infrastructure. The government and Eden District Council have done the right thing, therefore, to reinforce the Gilwilly estate with another £3 million for new road access. The same principle should make the government concentrate not on development grants but on broadband and mobile coverage, available to all. My father may not find it easy to describe what Britain does today. I don't either. But our mission should not be to hark back to a past of giant industry; but instead to seek to understand, or at least not harm, 5,000 companies, in a thousand different micro-sectors, offering employment which is strikingly humane and resilient, in one of the most beautiful parts of the world.

27 November 2010

My week in Parliament was dominated by back-to-back meetings about the sewage system, the Olympics, the Forestry Commission, army reforms and rural planning. In each, lobbyists cited little-known committees, policy wonks recited statistics, and everyone deployed acronyms. Yesterday, SME didn't mean Small and Medium-Sized Enterprise but Subject Matter Expert.

It took me ten minutes of repeated questioning before I understood that a charity which described itself as 'addressing urban regeneration and youth opportunity, through accountable and creative approaches to social inclusion' was, in fact, restoring a community hall. I was always trying to guess what was not being said.

Understanding Cumbrian economics is difficult in a different way. Last week, I visited the Innovia factory in Wigton. Having endlessly emphasised that Cumbria was about micro-businesses, I was now visiting a factory of over a thousand people that employed most of the men in the local town. Having emphasised that industrial policies in Cumbria almost always failed, here I was visiting a factory which had been first created in the 1920s when Jack Adams' development agency persuaded a London business to relocate to an abandoned site in Cumbria, in return for rent-free land. And which not only – contrary to my theory – survived, but flourished.

Dressed in a white coat, a hairnet and safety glasses, I moved between buildings which seemed fit for NASA. One was filled with fifty-foot waterfalls of moving viscose and a whiff of sulphur. A half-hour walk down a four-storey tower showed me the transformation of resin pellets into polymer film, mostly used for making labels on bottles. I misunderstood a lot. I thought that my guide said the polymer passed through a 'dye' and could not understand how colour defined the shape of the film. And even once I realised he had said 'die', I still struggled to visualise how a flat metal press could shape a bubble and why it was necessary to heat, cool and heat the same material again.

But the visit wasn't only a chemistry lesson, it was an economics lesson. Before I visited, I would have assumed that like other British industries of this kind, Innovia was doomed. Its energy costs were high. Ninety per cent of Innovia's film was exported but it was not even near a motorway, let alone an airport, so its transport costs were high. It relied on a thousand workers and engineers, and was competing with Asian countries with lower wages and a far larger pool of engineers.

Innovia had not made any of the technology, out-sourcing or productivity investments which I would have assumed were necessary. Rather than retiring old technology, the company had retained it. It had not out-sourced research. It had not dramatically cut staff numbers. It had not bought state-of-the-art computerised robots for cutting from the top international manufacturers. Instead, it retained a large research and development team in Wigton and continued to build much of its machinery in-house. Instead of retiring production lines, it had kept technologies – such as cellulose and bubble moulds – decades after competitors had abandoned them. They used people for tasks which others performed with computers and robots (more people were employed in cutting than in making the film itself). None of this was what we usually mean by technology, innovation and productivity.

Yet these decisions, which I would have predicted would have been Innovia's problems, seemed in fact to be the secret of its success. In the 1980s there were twenty cellophane plants in Europe. All stopped their viscose waterfalls and cellophane production – which they saw as a 1920s technology. Innovia stubbornly clung on to its cellulose and bubble moulds (at best a 1950s process). And then found itself in a world much more concerned about the environment, which suddenly again valued cellulose because it was biodegradable. And dealing with customers who wanted much thicker film, of the sort which could still be made with the 'old' technologies. There, human cutters still allowed them to adapt to different size orders far more cheaply and efficiently than a computer. As a result, they are now the world leader, or the second in the world, in every one of their products from polymer banknotes to transparent pressure-fixed labels (they make the Appletiser labels, for example). This is what makes them the largest, most profitable international export company in the constituency, and the life-blood of Wigton.

The largest threat they seem to face now is not from technology, globalisation or China but from the policies of the British government. The chief executive showed me that he had just been hit with a half-million-pound extra energy bill because of a lack of national investment in energy generation (the country as a whole needs hundreds of billions of energy investment). A cutter explained how the lack of bus services was making it impossibly difficult for his son to commute to the factory. Tom, who had started as an apprentice, confirmed that English schools are now producing too few good engineers for their research and development or machine manufacture facilities.

All these revelations during a couple of hours in Wigton are part of a pattern. In the last three months, I have sat in countless meetings in London, listening to the statistics and action plans from senior officials: trying to guess when they are misleading me or misleading themselves. Then I have visited Cumbrian businesses: solar panels and the haulage industry; milking machines; supermarkets and pub buy-outs; industrial parks and national parks. All challenge our country's industrial policy in different ways. Innovia revealed, perhaps more than any other company, how unfocused and contradictory my London-shaped assumptions have been about productivity, technology innovation, out-sourcing, education, the environment, the economy and energy. The problem is how to take the lessons of these very local encounters and hurl them like a grenade at the policy centre.

COMMUNITY LAND

CUMBRIA'S COMMUNITIES HAVE A LONG tradition of organising themselves. Two hundred years ago, Wordsworth found Cumbrian 'statesmen' (yeoman farmers) uniquely self-reliant, tough and independent-minded. They were in turn the descendants of the even more disturbing and egalitarian Norse farmers on their democratic and bluntly named 'thing-mounts'. The yeomen of Orton rejected and bought out their feudal lords more than four hundred years ago.

As I have said above, this was a heavily subsidised and regulated economy, but this had not created the 'culture of dependency' much feared by my Conservative colleagues. Instead, citizens often delivered services which elsewhere only government was expected to provide. By 2010, we had the only community-run snow-plough and ambulance in the UK. We had far more community fire engines, community broadband initiatives and community hospitals than any other part of the country; and more common land – 200,000 acres of it.

All these Cumbrian institutions – which had vanished in other parts of the country – were under immense strain from market pressures, central bureaucrats and government cuts. We seemed to be campaigning all the time, and I wrote about almost all of these campaigns. Many were designed to block things. Cumbria had probably resisted external imposition since the Romans built their new wall on top of their farms in AD 120. Wordsworth led the 1845 campaign to prevent a train line being run through the Lake District (he described the railway as 'a false utilitarian lure mid . . . paternal fields at random thrown').

Such campaigns are easily dismissed as 'not in my backyard' nimbyism, particularly today, when Britain's productivity, growth – and median wages – have been frozen since 2008, and when Britain finds it so difficult to build new railways, runways and housing. Our Cumbrian campaigns can seem the very epitome of what stops British growth. Janan Ganesh, a fair-minded journalist, captured this in the *Financial Times* when he wrote: 'Rory Stewart is a rural romantic . . . If you think it is hard to get things built in Britain now, be glad Stewart was never planning minister.'

Janan has a point. One of the *Herald* articles, not included here, describes my opposition to burying nuclear waste in Cumbria. But I did not always join my constituents in blocking developments. I championed placing hundreds of mobile telephone masts on hill-tops at a time when the majority of my constituents and many MPs were against them. I backed large concrete-and-glass flood barriers against the wishes of residents in Keswick. I argued for doubling the width of the A66 road, building a high-speed rail link between Leeds and Manchester, and for allowing more housing to be built beside small villages. I wrote about all these topics in the *Herald*, and if I haven't included the pieces here, it's because they are mostly dull, technical and worthy. (Maybe that's a clue to why the arguments for growth can be difficult.)

I have, however, kept two 'nimby' pieces on our campaign to keep wind turbines out of the Lake District and the Solway Moss. I have since changed my mind, and now think that I should have opposed the turbines only where there were strong historical, landscape, community and environmental arguments against them. I should not have opposed the Solway turbines.

I have kept both campaigns in, however, because the point of these letters is partly to show what I, as an MP, thought and did at the time, rather than what I ought to have thought.

Our community campaigns, however, were not restricted to nimbyism. There were also campaigns to save things – I led successful marches to save community hospitals and fire engines, along with failed campaigns to save police stations and magistrates' courts (one letter, included here, focuses on our campaign to save the Penrith cinema). And there were campaigns to get things which we had been denied: our largest and most successful effort – and the subject of many of the original articles in the *Herald* – was for superfast broadband (although I have dropped those letters here in exchange for one on affordable housing).

All the community campaigns raised different questions. How do you weigh up the beauty of a Lake District valley against the need for renewable energy? What business did we have trying to stop a private company from closing a cinema? (And why, if the company was right, is it still open fifteen years after it was supposed to close?) Why were we unable to replicate the success of community housing elsewhere? And was I – as an MP – ever more than a mascot and a follower in any of these initiatives?

These campaigns might seem small-scale to urban readers. But in sparsely populated areas, getting 150 people to a meeting represented skilful mobilisation. They required crowdfunding, social media campaigns, direct action, legal challenges, and an increasingly sophisticated understanding of planning law and local government finance. This was 'actually existing democracy' – not the version taught in schools, but the exhausting work of neighbours trying to solve problems when no one else will.

Together they raise certain fundamental questions about

how much say communities should be given. Could growth not arrive in a form that fitted local livelihoods, landscapes and structures? Could social capital, volunteer hours, traditions and resilience not become a source of long-term value – a value beyond GDP statistics? When is local resistance a type of unpaid due diligence: in effect pointing out environmental and community costs to planners who have overlooked these factors? When is loving your own landscape – or imposing infrastructure on someone else's landscape – selfish?

I begin with the question of the boundaries of the constituency – and what it means to create a coherent identity for parliamentary representation – and how history and prehistoric geology play into the modern choice. I end with a visit by the prime minister David Cameron, whose vision of a 'Big Society' seemed to find its most complete and rich fulfilment in Cumbrian initiatives, but who, confronted with the realisation of his dream, did not respond in quite the way I had expected.

17 September 2011

The Boundary Commission is a body appointed to review constituency boundaries. Unlike in the US, it is largely independent of political control. It does not, therefore, gerrymander. In 2010, it was given the unenviable job of reducing the number of constituencies from 650 to 600, partly because the ruling Conservative party had developed a spreadsheet which suggested this would favour them electorally. Their proposal would have torn in half the constituency I had tried to get to know, leaving my cottage in another constituency and giving me areas with which I had never interacted, and with whom I felt my constituents had little in common. In this next letter, I try to explain what can work when geology, and history, reinforce constituency identity.

The Boundary Commission has placed its chisel into the High Street ridge, where the Roman road falls to Troutbeck, and struck it with a hammer, shattering the county like a piece of Skiddaw slate. One long crack now runs towards the coast, another north to Carlisle, another south to Arnside, making a rent in our soil fifty miles long that splits Cumbria, its constituencies and its communities.

There are many more natural ways of dividing the county. The ridge-lines, the rivers and the watersheds separate the fellside from the lakes, the coast from the inland, the hills from the plain. The ancient boundaries of Cumberland and Westmorland follow those lines: and so too have our constituencies for nearly a thousand years.

The boundaries between the constituencies of Westmorland and Lonsdale, Barrow and Copeland, for example, are the exact boundaries of the older counties of Cumberland and Lancashire. The northern line of this constituency of Penrith and the Border was the exact frontier of the Wardens of the Western March in the early Middle Ages. Its other edges are all formed by mountain ridges – Helvellyn and Blencathra to the west, the Shap fells and the Howgills to the south, and to the east, the Pennine border, which was almost certainly the national boundary of the ancient kingdom of Cumbria.

The geography is so complete that from Wild Boar Fell at Mallerstang, or Hartside Pass above Alston, you can see the whole 1,200 square miles of the constituency and nothing else.

There are ways of creating natural communities within the Boundary Commission rules (the commission wants equal populations, and five, not six constituencies in Cumbria): you could group Whitehaven, Workington and Maryport; keep districts for Carlisle and Barrow, while still retaining two rural constituencies.

But this is not what has happened. In the hallucinatory scratchings of the Boundary Commission, Morecambe Bay, on the edge of Lancashire, is now to be combined with Nenthead, on the high Northumbrian border; and Windermere with Whitehaven, over Hardknott Pass. The central rent runs downwards, like a virtual wall, cutting the western edge of Carlisle, Penrith and Kendal, dividing central Cumbria's communities from their neighbours, and linking them with places with which they have little in common: Carlisle racecourse is torn from Carlisle and attached to Workington; Rheged is ripped from

Penrith, tied to Maryport. The Kendal bypass is transferred from Kendal to Whitehaven.

Does any of this matter in a global 'interconnected' world which can often make geography and cultural identity seem irrelevant?

Yes, because the ancient boundaries of Cumbria, which derive from geological and topographical reality, and run along watersheds, continue to separate quite distinct human constituencies with separate needs. Our limestone and volcanic hills gave us our record sparse population and our backbone of upland sheep farming. Copeland's coal-seams gave it first concentrated employment, then unemployment. Now it contains the radically different demands of Sellafield and the nuclear industry.

It is no surprise that I, therefore – as the Member for Penrith and the Border – ended up as chairman or officer of four All Party parliamentary groups: on mountain rescue, rural parishes, upland farming and rural services. Nor that the MP for Copeland chairs the work on nuclear fuel and is leading the campaign for storing nuclear waste in Cumbria, while I am active in the campaign against it. The separate geologies, populations, needs and claims of our constituencies should not be smudged and elided, but instead represented crisply and distinctly to policy-makers in London, who have little time or patience, or experience of either the nuclear industry or upland farming.

It is not always easy to get things done in a parliament where many MPs are indifferent to our needs, and where we should sometimes challenge the mass, momentum and whipping of the political parties. If, therefore, Cumbria's MPs have succeeded in defending Cumbria's corner, it is generally because we have specialised not just in Cumbria but in our specific constituency in Cumbria. The West Cumberland Hospital was saved by West Cumbrian MPs; but it was the Cumbrian MPs in the Lakes and fells who fought for the uplands, for the community hospitals of Penrith and Alston, and for farmers in the face of foot-and-mouth. It will dissipate that focus if a colleague – whose experience, vocation, voters and inclination have led him to fight against urban poverty in Whitehaven – now finds themself responsible also for the hiking shops of Ambleside. In the new proposed Workington constituency, the needs of the different communities, which are to be jammed into one, are so radically different that life expectancy will be twenty years longer in the east of the constituency (Greystoke) than it is in the west (Maryport).

It is not policy papers but visits to constituency farms which teach MPs the problems of under-stocking on fells; it is a neighbour with Parkinson's, not a YouTube presentation, which convinces an MP to fight for broadband video links between rural hospitals. And the reason why we have a voting system that relies on constituencies defined by local areas (rather than proportional representation) is precisely to amplify the spirit of place and to allow local areas their own influence in Parliament.

MPs work well when constituencies work; and constituencies work when communities act together. The Boundary Commission's new proposal to create severed stumps, sewn together with coarse thread, will spawn doubly divided monsters: lumping separate communities, and separating unified communities. While, in our different ways, on different issues, Cumbrian valleys should echo with separate and distinctive voices.

The 2010 proposals from the boundary commissioners were rejected because Parliament failed to pass the bill reducing the number of seats.

My constituency retained its boundary throughout the ten years in which I represented it. Three years after I stepped down, however, a new proposal came forward, driven this time by an understandable wish to make constituencies across the country more equal in population (my constituency had 70,000 voters, London seats had 100,000). The result was to tear the constituency apart in a quite new direction.

Penrith and the Border was no more.

22 January 2011

I spent Christmas with my family. My little sister, who has Down's syndrome, made Christmas for us. She spent much of the holiday in a pink, pointed woollen hat, with ear flaps of the kind I imagine a Peruvian wearing. She took my father swimming and me skating (I promptly fell over in the middle of the ice: she didn't). She took my mother 'pottering', which seemed to mean walking round the post office (I don't know where she gets the words from). Her delight in presents, the methodical pace at which she opens them, her strong sense of ritual – telling us when we are allowed to eat cake, when to sing carols, her care for cooking – imposed a discipline and tranquillity on us all.

I spent the last ten days of the recess alone, trying to finish some writing, while three hundred miles away the political wheels in London were creaking into action with scandals, open letters and rumours of reshuffles. I didn't see anyone on New Year's Eve but the next afternoon I walked out to Askham, climbing from the ruins by the beck to the pit burials on the fell, turned at the standing stones, and came back along the Roman road over High Street. Otherwise, I was mostly dealing with correspondence.

Many emails protested against the actions of powerful companies. Developers were forcing housing estates into valleys, wind turbines onto mountains, and supermarkets into market towns against the wishes of communities. The same supermarkets were crushing dairy farmers by paying an incredibly low price for milk. All that is local, from the Penrith cinema and Wigton's shops to Newton Rigg, seemed to exist under the shadow of big, national organisations with short attention spans.

The letters and campaigns against these developments were almost always compassionate, knowledgeable and determined. Parishes or groups of parishes like the Upper Eden Community Plan displayed an imagination and generosity which would be difficult to simulate in distant offices. And communities care about results: Lyvennet protects its landscape because it loves it; Crosby presses for affordable housing because they need housing for their children. And a hundred brains working together generate remarkable solutions.

In Greystoke, in Patterdale, in Mungrisdale, in Alston, in Mallerstang, in Bewcastle and in Morland, communities pressing for broadband – researching, publishing on their website, suggesting new technological and engineering solutions and circumventing legal and financial barriers – are producing solutions which are quicker, smarter and cheaper than any previous government initiative. From Hartley to Penton, lecturers, senior civil servants, lords, farmers, councillors and former head teachers have produced serious and considered proposals for Newton Rigg, without which the site would probably already have been closed. It was the communities that saved our community hospitals. And the hundreds who gathered last week to march for the cinema will with luck and a great deal of effort be able to save that, too.

But the effort required is immense. In each case communities are taking on giant organisations, public or private, who own the sites, have the financial information, have rigid theories about what should be done, and who have often out-manoeuvred other communities in the past.

Communities must struggle to find out basic information (such as the real size of Newton Rigg's deficit); study laws and restrictions (such as whether European 'state-aid' rules really stop them using public fibre); challenge technological claims (such as how much energy the wind turbine generates); they must relentlessly press officials to act.

This is exhausting for communities: it risks their time and money, and in the process, exposes them to their opponents' accusations that they are ignorant and small-minded. But if communities are trusted, then the compassion, the knowledge and the will-power which they can bring to bear is irresistible.

The spirit extends further. Last week, I had finished my *Herald* article and sat by the fire trying to work out how to defrost my meal (the Aga and the heating had been broken for a week and the electricity had gone off again) when there was a sound like a trapped jackdaw clambering up, slipping down and struggling in the chimney. Then black clods began falling onto the hearth and an acrid smell, like burning yak dung, filled the room. Orange flames appeared behind the mantel-shelf. I pulled off the mantelshelf – revealing for the first time since 1900 three sepia prints of ladies in Victorian hats, a set of scissors, a penknife and a live pistol round – and poured buckets down the wall.

I was saved by the Shap fire brigade. The firefighters had been dragged out of sleep at one in the morning, had located my cottage on a nameless track which was invisible to any GPS, and had then had to abandon their engine and come in half a mile on foot on a narrow icy track in driving rain. But they were as cheerful and calm as though it were a Sunday-afternoon picnic.

One firefighter worked the thermal camera, another pair

assembled a pump which seemed to include a pipe, partly made of Victorian bamboo, attached to my watering can. They had the situation under control so quickly that I had nothing to do other than to walk up and down the hill to carry some salt to help their fire engine off the track. Here was the same combination of compassion and energy which we see in communities – and a reminder that the virtues of villages and charities can also be found in the public sector.

24 November 2012

In many Cumbrian villages, residents cannot afford to buy or rent homes, so they leave, taking their families and their businesses with them. As a result, shops, pubs and primary schools close, and villages become increasingly reserves for the elderly, whose children and grandchildren live in distant towns.

We talk about this all the time. But what do we do about it? How do we produce houses which the young can afford to rent or buy? In earlier periods, councils built the houses themselves. Sometimes, as in Penrith in the 1930s, this resulted in beautiful, well-conceived estates, still treasured today. Sometimes, as in Wigton, it resulted in places which have some unruly fringes. Sometimes, as in Bewcastle, the houses were so remote that the Carlisle residents assigned to them refused to stay. But there is in any case no longer the money for councils to build the houses themselves. The new idea from policy-makers is to allow developers to build full-price houses and then require them to use some of their profits to subsidise a dozen affordable homes. This, however, means that affordable housing depends on giving permission for large numbers of unaffordable and often unpopular new developments, which make villages uglier and bigger.

Crosby Ravensworth – a picture-perfect village whose name implies a blend of Christian and Viking, and whose gentle stream runs past gleaming dry-stone walls and a Norman church – has produced a different solution. Its cluster of whitewashed houses seems the epitome of an ancient Cumbrian community. Except no one can afford to live there any more. The average house price was £315,000 in 2010 – eleven times the average household income. The shops and the primary schools are on the verge of collapse. The last pub had shut its doors. Eight in ten residents had not been born in Crosby. And a dozen families who worked in, lived in, or had connections to the village, couldn't afford to rent there.

They didn't want a developer building a hundred homes on a greenfield site. So they decided to build themselves. They identified a good place on the site of an old stone business in the village centre. They wanted to build twenty-two houses, rather than trying to squeeze in the thirty-four which the planners insisted should fit. And they wanted the affordable houses to be larger, more attractive, and better designed than the standard.

There must have been many occasions when they wondered why they had ever begun. Their work had all the intensity, risk and personal responsibility of setting up a small company. Many people in the village worked unpaid for two years, putting in all their spare time after work. They did it not for themselves, but because they understood how important it was to keep young people in their community.

They learned acronyms they never wanted to hear, encountered agencies they never suspected existed, and were shuffled from architects to code assessors, from engineering designers to surveyors and builders. They were drawn into the strange world of grant proposal writing, agreed to be something called a Big Society vanguard, and struggled with the sustainable building code.

They received a grant from Eden District Council, and one from the Homes and Communities Agency, and borrowed over a million from a charity bank. They were nearly stopped by the discovery of rare bats, and it seemed for a moment as though the money would never come, and the entire project would collapse. And by the end, one wondered how they had the energy to continue.

But they succeeded, astonishingly quickly. Because they did it themselves, there were none of the objections which you find when development is imposed from outside. This allowed them to complete far more quickly than a commercial developer. A year and a half ago, when they began, there was nothing to be seen in the centre of the village except cracked concrete paving stones and the bat-haunted quarry sheds. Three weeks ago, we buried a time-capsule (containing a copy of the *Herald*) in the grass of a new village green. Around us were twelve new homes, all affordable and available to be rented or part-owned. They were arranged in a square, with projecting wings and slate roofs, some rendered, some faced in limestone and some in sandstone: a very Cumbrian concoction sitting comfortably in the heart of the village, without any two houses quite alike.

Between them you could glimpse on that sunny autumn morning sheep and fells, the community hall, the church. The houses were owned by the village, in a community land trust. Behind them was the pub, also saved by the village, also owned in common.

The houses are not twee. But nor are they expensively individual. All the designs are made from just two standard kits – one for a two-bedroomed and one for a three-bedroomed house – arranged in different combinations and facings.

They are affordable to build as well as rent. There are

broadband ducts ready in every property. The houses are heated by air-source heat pumps, with no oil or gas. The residents pay an electricity bill of only £7 or £8 a week (their neighbours pay £90 for the same services). The land trust has ensured the houses are limited to locals in need. There are now tenants in all ten of the rented affordable houses, with eight people under eighteen. And a brand-new addition was born this Monday.

Now a dozen other villages – Culgaith or Lazonby, Barton or King's Meaburn – could, I think, do the same. Some things will be easier second time round, and the pain of Crosby Ravensworth may save some pain for others. Extraordinary figures like Andy Lloyd, of the Cumbria Rural Housing Trust, can help establish community land trusts. And Crosby Ravensworth has offered to share its experience.

Some things will be more difficult: there will probably be fewer grants available, and communities will need help securing larger loans. A community must still put immense time and effort into developing the kind and number of houses it wants, for the people it wants, in the place they want. The government and charitable foundations need to be more flexible, and imaginative, in supporting such schemes. We all have a lot to learn before we can spread this model across rural Britain. But Cumbria, and in particular Crosby Ravensworth, has proved magnificently what can be done.

Crosby Ravensworth's startling and impressive achievement seemed to be a model for a dozen other villages. None followed, partly for the reasons I worried about in the paragraph above. But largely, I felt, because they did not have the individual leadership provided by people like David Graham, who led much of the work in Crosby Ravensworth. They wanted affordable housing. But they preferred to wait for the

4

government to build it, just as they demanded council action on their roads but did not want to adopt Alston's model of a community snow-plough. In neither case did the state oblige. The East Fellside villages remained without affordable housing and their roads were often blocked by snow.

9 July 2011

The small is displaced by the large, day after day: the smaller hill farm by the larger; the market store by the supermarket; the community by the district hospital. The same story has been repeated over the last fifteen years with schools and post offices, dairy farms and pubs. And it is happening now to our Cumbrian charities.

More people are involved in voluntary activity in Penrith and the Border than in almost any other constituency in Britain. In the last few weeks, I have visited Greystoke's Sunbeams Music Trust, Wigton's Chrysalis, Bewcastle's Low Luckens Farm, Brampton's Community Trust, Carlisle's Eden Valley Hospice, Heathland's Glenmore Trust and Appleby Heritage Centre.

One of the charities consists of a husband and wife and twenty-seven acres; another has four hundred volunteers. One receives half its funding from the government; another raises £7,000 of private donations a week. One is in a purpose-built building with bright modern furniture, another in a set of railway carriages. Some deal with eight-year-olds; some with eighty-eight-year-olds; some with housing; others with health. But each charity was born in Cumbria, and all still bear the mark of their origin as a tiny charity, and their founders' desire to meet a particular Cumbrian need.

Hospice at Home, for example, was created in 1987 by four volunteers who realised that 80 per cent of the people who died in hospital would rather die at home. It has survived all the tumult of NHS reforms, all the risks of growth and all the funding crises, to now serve hundreds of patients a year, with more than 10,000 hours of care. It has staged grand fund-raising events (I still haven't lived down dancing to Lady Gaga with 150 Santas in Penrith) and it has been honoured with medals at Buckingham Palace. But its stall is still at Skelton show, and Fiona, who was a founder of the local branch fifteen years ago, mans that stall. Like almost every other Cumbrian charity, Hospice at Home still relies largely on unpaid volunteers to reinforce a few effective and experienced full-time staff. It keeps its overheads very low. It does not employ large teams of 'development professionals' to write grant applications.

Now giant 'third sector organisations' are competing against them. These big national charities have professional management, and economies of scale. They also have skills and good track records, but this is not why they consistently win contracts: they generally win because they are richer. They have revenues of hundreds of millions of pounds a year: a hundred times the size of the largest Cumbrian charities. Some employ more than a hundred people just in their fund-raising and 'development' teams.

And they know how to promote themselves: this week alone, I saw ten events organised by big national charities in Parliament, many with celebrity guests, videos and pink-iced cakes. Yesterday, I was presenting an award at a London ceremony for charities that had so many glamorous assistants, spotlights, sports stars and tracks of ethnic drumming that I thought I was at an American ball game.

But it is not the cakes which make big charities so appealing to foundations and government: it is that they feed our new obsessions with 'professionalism', 'accountability', 'risk-assessment' and 'sustainability', all of which are more easily demonstrated by large teams of professionals. Contracts are increasing in scale. Application forms demand elaborate strategies and needs assessments. Insurance costs are rocketing (as even Crosby Ravensworth village fair is finding to its cost). Compliance favours the most exhaustive monitoring and documentation. The most straightforward act of neighbourliness now seems to require a training course. It is much easier for the government, therefore, to give the contract to a national charity, with its financial base, sophisticated reporting systems, right qualifications and right words.

But it is often worse for the 'service users': the Cumbrian with learning needs, or the terminal cancer patient. Local charities have local knowledge, leadership and flexibility, and are not the prisoners of a complex national template. What they lack in formal strategic plans, they produce through human relationships. There are fewer qualifications but often more local compassion and understanding.

Take Eden Mencap. Once or twice a week, a friend of mine, with severe learning difficulties, walks into the office which has been on Penrith high street since he was a child. There, he is supported by people who have known him almost all his life – often just by talking to him, but last week by gently unpicking the hundreds of pounds of debt he had run up on his Argos card. But a national contractor might ask whether it was efficient to keep an office on the high street. They might determine that my friend is not in the correct category to receive the type of government support which they are contracted to provide. Eden Mencap does not ask those kinds of questions, and therefore my friend and hundreds like him have been able to rely on them for subtle, difficult-to-categorise needs for twenty years.

I am now working with Cumbria Voluntary Service to bring small local charities together with donors. The hope is to create simpler processes and more manageable contracts which reward local knowledge and experience, rather than hyper-polished proposals. I've asked Whitehall for ideas on how officers can adapt the procurement regulations. But the big national charities could also help by showing more sensitivity.

Some already do. Last week, I was at the wonderful new Roman Frontier Gallery at Tullie House, Carlisle, with the director of the British Museum. The British Museum, like the Liverpool Tate gallery, could have set up a 'BM North'. It has the money, reputation and the masterpieces to succeed and take visitors and funding away from local museums. Instead, it has chosen to lend its treasures to Tullie House, under the banner of the existing local museum, and work with its curators to build on what is already there. I hope more of the great national charities will follow its example.

19 February 2011

My opposition to wind turbines used to be merely theoretical. I understood that they were an inefficient way of generating energy and that companies pushed them into inappropriate places, against the wishes of communities, because of extravagant subsidies. But friends told me I was out of date, that I was not taking the environmental problems seriously, and that Britain desperately needed clean energy. I also knew struggling farmers, for whom turbines could bring enough income to save a family farm. So I was confused.

I am no longer. My visit to Reagill has turned my theoretical opposition into something much deeper. Penrith and the Border should be a turbine-free constituency. Standing above Reagill, you can see the horizon from Blencathra to the Pennines. Nothing breaks that silhouette and, although it is the largest constituency in England, you can see almost every mile of it: an unbroken sweep of rolling hills, clean to the mountain ridge-line, without industrial machinery. It is almost the last constituency in Britain where this is true.

In Scotland, the crest of the wild and empty hill of bracken above Glendevon is visible for 3,000 square miles around, from the Sma' Glen to the Pentlands, south across the Forth: and for

3,000 square miles around you can see that it is crowned by giant white turbines.

We are a densely populated island that industrialised heavily and early, and has no Alps, no wilderness, no Himalayas. Since Wordsworth, the British have found their escape and freedom in our Cumbrian landscape. People still want to move and retire here; live and die here because of the purity of the space. It is a human-made landscape, sculpted and tended by farmers, but the human hand is quiet and restrained. It consists of pure expanses of open ridge-lines framed by what we call mountains, but the hills are, in reality, small in scale.

As Auden says, this is a limestone landscape 'Where everything can be touched or reached by walking'. If 400-foot industrial turbines taller than St Paul's Cathedral whirr on 300-foot hills, sending flickering shadows over the valleys, all this is altered forever. Everything around will seem suddenly domesticated, humiliated and diminished. The hills will seem, under these great lazy robots, like municipal parks in the shadow of a Ferris wheel.

So what? During the election my opponent and friend said to me: 'You can't pay the bills with the view, my lad.' Turbines seem to promise great environmental and financial benefits; they are astonishing technology and can be starkly beautiful; they epitomise modernity. Carbon dioxide emissions are driving climate change and could destroy the livelihoods of billions: we are not even beginning to properly reduce emissions. Our constituency has some great renewable projects: geothermal in Gamblesby, solar in Bolton, wood in Alston, hydro at Bongate weir, but they do not begin to suffice. Even the contribution and sacrifice that Cumbria has made in the nuclear generation is not enough to solve Britain's problems. The arguments for wind turbines seem hard-headed, practical and moral. Are we, therefore, simply being foolish, sentimental, selfish?

No: because if we kill the magic of Cumbria by driving 400-foot steel stakes into the heart of our landscape, we lose more than sentiment. For a start, we will lose money. Tourism is the largest income earner and employer in the constituency, and the landscape, which we will be wrecking, is what the tourists pay to see. They come now to visit one of the last upland areas in which it is possible to see how Britain looked before the wind turbines.

But what moved me most on Saturday above Reagill was not simply the beauty of the hills, or the importance of protecting our landscape for our 'visitor economy'. I was moved most by the fierce commitment of the anti-wind-farm campaigners themselves. Once you have met them, grasped the time they have given, the research they have done, the opportunities they have passed up, the way they live their lives, you would not call them nimbys. When they talk about the skyline, or point to the birds on Winter Tarn, or discuss their support for the fights of other communities from Shap to Stainmore, they are neither sentimental nor selfish. They are people whose lives are concretely absorbed in their landscape: aware of its shape, its heft, its space. They love it as millions of others before them have loved it, and as millions after their deaths could love it.

11 June 2011

We were nervous about the rally against wind turbines last Saturday. We worried that nobody would turn up, or that a crowd of hostile 'antis' would be bussed in to shout us down. But by 11.30 a.m., fifty yards north of the Scottish border, there were more than two hundred people, well wrapped against the cold.

Two dairy cows were reaching hungrily over a fence towards a cardboard placard depicting, in felt-tip, a turbine, but there were no 'antis'. A huge blond man in a black suit, from BBC Newcastle, was giving frantic instructions to his cameraman. He had the expression of someone struggling to record some outdated and troubling ritual involving tweed-suited conservatives, or even perhaps druids. But he would have struggled to stereotype our group.

He was between a nurse from Longtown, some sheep farmers, a man who repairs tractors and the heads of the parish councils from Mungrisdale and Tebay. Some were in ties, some in tracksuits. A retired colonel stood next to a young man who had the stubble and clothes of a rock musician. There were children in bright wellies, a man in his forties with a ponytail, and a bishop.

They represented communities from over 2,000 square miles

of Cumbria: from Wigton to the west coast, from Lunesdale to Longtown, from Shap to the south of Scotland.

The protest was against a proposal to build nine turbines on the Solway Moss. We tried to explain to a reporter how this flat, open marsh land would be altered by nine spinning structures, each four hundred feet high. We pointed him to the highest man-made object in the landscape – a telegraph pole – and explained that these turbines would be not twice, but twenty times that height. But it's hard to imagine, until you've seen it, a white steel-and-plastic turbine, turning in the evening light and visible for four hundred square miles around. The first Scottish speaker spoke of how it felt to find herself living under the flickering shadow of such giants in her remote Border valley: she is soon to have more than four hundred in her parish alone.

We make more than a billion pounds a year from tourism: it is our largest income earner and supports more than 10,000 families in Penrith and the Border. We are not a wealthy area and hill farming has had a very difficult time. And why, I asked in my speech, do tourists come here? We have great meat, but they are more likely to go to France for the boeuf bourguignon; to Venice for Gothic palaces; and if they want sun, they go to Spain.

Greece is almost as convenient as Hadrian's Wall if you're coming from London. Our greatest resource is not our wind but our landscape, and turbines will destroy it.

Nor will building them in the northern Lake District save the world. Converting coal power stations to gas, and developing cleaner cars, would have 10,000 times more impact on reducing the UK's carbon emissions than wrecking the Solway Moss. The new discoveries of shale gas in the United States and Europe answer many of the concerns people had, even eighteen months ago, about energy supply and energy security. (Professor Dieter Helm, of Oxford University, has written powerfully about this.)

And Cumbria is already doing an enormous amount to generate non-carbon-emitting energy. We have just agreed to build three new nuclear power stations which most parts of Britain would refuse, and with them generate 3,600MW for the nation. So why should we now also become the national dumping ground for wind turbines, when all the turbines proposed for Penrith and the Border would generate 1 per cent of the power we will produce from our new nuclear stations and wreck our landscape, our homes and our economy in the process?

A hundred of us, therefore, moved on that afternoon to Longtown Memorial Hall community centre to discuss how to ensure Penrith and the Border be a turbine-free constituency. Dr Mike Hall gave a brilliant presentation on energy and economics. Adrian Todd told a group on the Scottish border how Tebay had gathered its environmental information; Bewcastle explained, from painful experience, which arguments to avoid; Wigton asked Scotland about techniques for collecting noise data; Berrier Hill told Longtown how to raise money for a public inquiry. I set up a basic website to record and share this knowledge and make it available to other groups. Our hope is that the turbine developers will begin to recognise the depth and strength of the opposition, the importance of our landscape to our economy and our lives, and will stop trying to force such developments through.

Cumbria is not being selfish. I knew enough of the people there to know that they were the exact opposite of 'not in my back as 118 yard' nimbys. The campaigners were some of the most generous and active members of our communities. Many were there without any personal stake. Some lived in a national park, where turbines are banned; others had fought and won their cases against wind turbines long since, and had driven north to join a protest at a place fifty miles from their homes. They were there because they cared deeply about our Cumbrian landscape, and about other communities who find meaning and solace each in their own particular landscape.

The application for wind turbines in the Lake District was defeated but the Solway turbines were built. I now feel that the planners were probably right and I was wrong about the Solway turbines – although this would be difficult to explain to my constituents on the Solway.

5 February 2011

On Thursday, I defended Cumbrian community action in the London School of Economics, where the rhetoric was as extreme as the setting was bland. My opponents, enraged by the prime minister's talk of 'Big Society', compared parish council initiatives to Athenian slavery and the bombing of Baghdad – all in front of a quiet scholarly audience in a lecture theatre.

On Saturday, back in Penrith, I found myself amidst the joyful and unsettling reality of a demonstration to save the Penrith cinema. I teetered on a small stage above a crowd of perhaps six hundred. Some protesters stood beside the police, holding banners, as solemn as at a picket-line. The group by the toy shop were also funereally quiet; but then from the Co-op came a sprinting, whooping posse in neon bodysuits, leaping on the stage like circus tumblers. Children were singing and old men had that astonished, happy expression which I associate with a successful wedding.

When I invited people to buy a share in the cinema, fifty voices shouted 'yes' in unison and we raised £9,000 in nine minutes. Dawn, who was leading the march, dressed for the second week as Darth Vader and chanting 'Save our cinema!', seemed to harness the shared faith of the crowd and its incongruous insurrectionary energy.

The next day, I was worried that we had all been simply carried away into something very risky. I had helped convince Graves, the owner, to let us bid, but was the price right? How was this whole thing going to be managed? I cleared my Sunday afternoon and arrived in the Methodist Central Hall, expecting a meeting of half a dozen. There were forty: some in jeans and weekend stubble; some in climbing gear, fresh from the hills; others dressed for the 'Sunday Evening Extra'.

It was a group of varied equals, from teenagers to pensioners: each with firm and different philosophies and political views. But Ruth, whom I had never met before, somehow emerged as first among these equals. She ran the meeting with the crispness of a conference in *The West Wing*: assigning every task from structural surveying to marketing, to share issues and charitable structures. She used jargon phrases – she talked about 'task and finish groups' – to which I have always been allergic, but she used them accurately and to deadly practical effect. I took on a small amount: setting up the website, finding an office space, contacting the seller. A teenager took over running the website, a council officer brought clarity and systems without ever imposing his views on the room, an accountant volunteered for the book-keeping. We've a very long way to go. But it's difficult not to be optimistic when you see the people in that room: 6,000 people have already signed up to support the cinema.

We now need an equally crisp, clear, single approach to saving our agricultural college at Newton Rigg. It has a historic name, unique land and strong support in the county. It could flourish by investing in new agricultural technologies, in dairy and in upland farming. And we need Newton Rigg for our economy, for our students, and for our farms. But it has been at imminent risk of closure for eighteen months.

There has been no lack of community support: Ann Risman and members of the Applefell group immediately launched a campaign, produced a compelling plan for the future of the campus and relentlessly kept the subject alive. The *Herald* has championed it. Eden District Council crafted a detailed Cumbrian bid to take over the services. The staff brought their own proposal. And there have been leaflets, demonstrations and dozens of official meetings. But harnessing local energy is easier if it is simply a question of persuading the cinema owner to keep the doors open. In this case, a very detailed plan for land ownership is required.

The proposal favoured by the government is to hand the college to the Yorkshire agricultural college, Askham Bryan: they have good skills, experience, a good plan and money. But this does not mean that Newton Rigg will be safe. Because Askham Bryan has demanded as their price all one thousand acres of Newton Rigg land – land worth perhaps £10 million – which was given by farmers and landowners, free to the county, for an agricultural college, a hundred years ago. Askham Bryan want full ownership with no strings attached. And if things go wrong, they would simply sell the land, take the proceeds and return to Yorkshire.

This would not just be the end of the land. It would be the end of over a hundred years of history, the end of hundreds of jobs, and the end of agricultural education in the uplands and the north-west. We need one last push in this long, difficult campaign, and one focus: to ensure Newton Rigg is a flourishing agricultural college in a generation's time. We must insist that, whoever takes the college, the land is protected by trust or strict covenant.

There will be no movie heroes in this fight: the final terms will be set by accountants and civil servants, and above all lawyers. But we still need some of that cinema energy for Newton Rigg.

As this letter explains briefly, Askham Bryan, a Yorkshire college, had volunteered to take over Newton Rigg. But they wanted as their price to take ownership of all Newton Rigg's valuable farmland. I and others feared that they would eventually abandon the college, sell the land and take the money back to Yorkshire.

Following this letter, I said that I would only support the bid if Askham Bryan left the land in a trust. The director agreed to do this and shook on the deal in the Pugin Room of the House of Commons. But she then reneged and took over the college with the land. I boycotted the new entity, but I was finally persuaded when five years had passed to visit and support again.

We managed to keep Newton Rigg open for another ten years. In 2019, there were still almost nine hundred students studying, focused on everything from farming to gamekeeping, forestry and horse management.

The year after I stepped down as Member of Parliament for Penrith and the Border, however, Askham Bryan announced it was closing the site and selling the farmland. A commercial bid, apparently including the factory which rendered animal fat beside Penrith, bought much of the land, which had been donated to the college. The students and teachers departed.

The cinema remains open.

20 August 2011

The last time David Cameron came to Penrith and the Border, it was forty-eight hours before the 2010 election. Just before eleven o'clock at night, I was outside the Border Cod in Longtown. And then, in an instant, young special advisers were spilling from hire cars; plain-clothes policemen materialised behind bus stops, ear-pieces crackling; and a giant purple 'battle bus' hove into view.

Cameron came out into an explosion of camera flashes, weaving between tripods and microphones. We managed only three sentences before he had collected his fish and worked his way back through the press onto the bus, leaving me, and the gathering Longtown crowd, gazing at the high, dark windows of the coach.

I tried to talk to a group that came out of the Graham Arms. 'We don't want to talk to you,' they said. 'We want to talk to him.' Since then, I've been trying to get him back for longer.

Our problems in Cumbria are the problems of his constituency writ large. Cameron has 400 square miles and 100 hamlets and villages; we have 1,200 square miles and 300 hamlets and villages. His is the largest constituency in the south; ours, the largest in England. He has been campaigning to save his village's pub, pushing for faster rural broadband, and notoriously to implement more 'Big Society' initiatives.

All of which – I keep telling him in informal requests and what seem innumerable conversations over cheese straws from Liverpool to Number 10 – means that he needs to see Penrith and the Border.

Last week, I reminded his chief of staff that Crosby Ravensworth had laid the foundation of the new affordable houses, and that the refurbished Butchers Arms (which hundreds of us had banded together to buy) was about to be finished this Wednesday. But the PM was dealing with the riots, and he has 650 other constituencies to think about.

So I was surprised this Monday, when I was on an official visit to Washington DC, to get an email in which the PM wondered whether he might turn up in Crosby in two days' time.

He wanted to be there for the opening of the pub, hear about the affordable housing, visit the new National Citizen Service pilot at the Outward Bound centre at Howtown, see Carlisle and, if he could, stay overnight and fit in a swim in Ullswater.

I cancelled my meetings, booked a new flight from the States and picked up the phone to Crosby Ravensworth, and Catherine, in the office, worked through the night to arrange all the visits.

I was not allowed for security reasons to tell anyone that the prime minister was coming. When I suggested to Joan Raine, the chair of the parish council that 'a senior visitor' might stop by on Wednesday, I heard that everyone from Crosby was due to be at a race meet at Carlisle. And when I wondered if the senior visitor could open the pub, David Graham, who drove much of the affordable housing, said: 'Absolutely not: it's a community project, and will be opened by the community. I don't give a monkey's who the visitor is. I don't care if it's the prime minister.'

I drove straight from my last New York meeting to the

airport, flew through the night, raced to Euston, just made the train to Oxenholme and arrived in Crosby, still in my crumpled suit, ten minutes before the PM. David and the others from Crosby arrived with a ploughman's lunch made by the Crosby Ravensworth Food Alliance. While the PM focused on his plum pie – he said he had failed 'disastrously' to make plum pie that weekend – Libby told him about digging broadband trenches in Mallerstang, and David explained how the community had begun twenty-two affordable houses across the road, Tom described the new neighbourhood plan for Upper Eden, and Cameron Smith and Kitty explained how we had put together almost £300,000 to buy the pub. Gordon Nicolson unveiled his new housing plans, and showed how Eden District Council had cut its deficit.

I saw the visit as a way of thanking the prime minister. These projects had come about because his government had backed our bids to be national pilots. We won support for neighbourhood planning, housing and the pub because he made us a Big Society vanguard. Then, as 'a broadband pilot', we had received more money per head for broadband than almost any county in England; and finally we had been made the pilot for National Citizen Service, which is why he and I built and launched a raft with students on Ullswater later that afternoon.

This was his chance to actually see what he had supported: to meet the people, to look at the foundation stones, smell the fresh paint, and grasp how much had been achieved in a year, and how!

I don't think he'll forget David Graham's short speech: 'I just wanted to remind you: you're not opening the pub but we'll let you open the bar'; nor will he forget standing with me up to our knees in Ullswater, hoping that our 'lorryman's hitch' was going to hold the raft together and stop the children falling into the lake.

I won't forget his smile as he looked across Helvellyn in the afternoon light. And I'm confident next time I assail him over a cheese straw, he won't forget Cumbria either.

At the end of the visit, I asked if I could join the prime minister in his car as he was driven to his hotel and he nodded. I got in the back seat beside him, and we set off with his security convoy behind in a knot of hefty black vehicles.

I remarked that this was the largest, most sparsely populated constituency in England.

'I represent a very rural constituency too,' he said sharply, as though he felt I was lecturing him.

'Yes . . .' I did not point out that Whitney was about a quarter of the size of Penrith and the Border, 'but,' I continued, 'we do have some very particular needs. The volunteer ambulance, up there on Alston ridge, for example . . .'

He looked out of the window at the fell. I mentioned that he was staying only five miles from my cottage. I wondered if he wanted to meet up later and if I could show him anything. He said he preferred to have a quiet night on his own. After a pause, he asked me where I had been for the summer, I said that I had been briefly in the States and before that in Afghanistan. He didn't ask about Afghanistan. But he asked where I was going next, and I said that, since it felt like Tripoli was about to fall, I was going to Libya and offered to brief him when I returned.

'Try not to get yourself killed,' he said, not commenting on the offer of the briefing. 'We wouldn't want a by-election.'

Tripoli, 3 March 2012

Last summer, the Corinthia Hotel in Tripoli was filled with reporters and photographers. They had propped their laptops on tiny marble tables in the lobby. Waiters brought Turkish coffees, but the reporters' eyes flicked only from their screens to their phones, checking for messages about Gaddafi's whereabouts, a recently discovered palace or prison, or a press conference. Only Marie Colvin seemed to look around the room.

A tall, elderly Libyan man had driven me to the hotel in a battered Japanese car, so small that he seemed barely to fit under the steering wheel. In the hotel, he studied the chaos politely. The new minister of the interior had just entered, flanked by recently promoted policemen, and the minister of finance had just left, grinning, in a phalanx of eager, American-educated advisers.

Some of the reporters thought they should interview the ministers, but nobody was interested in my taxi-driver: an elderly former water engineer. But Marie asked us both to join her. Since she was the journalist who knew Libya best – the only one who had known Gaddafi – she might have felt she didn't need to bother with my companion. Instead, she took a patient, courteous interest just as I had seen her do in other countries, when everyone else was too busy.

In our hotel in Iraq, in 2003, she had been the only person to talk to the hotel pianist. She discovered that he had been in the national orchestra. 'I used to play with the greatest musicians in the world,' he would tell her. 'Now I am in a hotel bar, and all I get to play is "Feelings . . . Feelings . . ."' Eight years later nobody could remember his name, except Marie.

On this occasion, my driver said nothing that seemed remarkable to me: he repeated that his country was a peaceful place, and that now Gaddafi had gone, everything would be fine. Piqued by Marie's interest, other younger journalists shuffled over. A couple asked questions, but they could get little from him, and they moved on. Only Marie sat patiently, took notes, thanked him warmly, took his number and promised to be in touch.

The other correspondents seemed worn down by weeks of reporting. Most didn't speak Arabic, so they needed a translator; colleagues were being kidnapped, so they needed a sensible driver; and the only place where they could get access to the internet (to file their stories), or indeed basic security or a few hours of running water, was this five-star hotel which might have reduced its sewerage but not its prices. Everyone was short of money. And everyone was being chased by editors to beat each other to the same stories. People were not going out of their way to help each other.

Marie, however, had an adopted family: a Libyan woman who was staying in her room, a Libyan man who was borrowing her laptop, and two young English stringers from the *Telegraph* and the *Independent* whom she had offered to take along with her to share her interview with a minister. She wondered whether I needed a shower, and lent me her satellite phone to call home. Satellite calls are very expensive, so I made a quick call and handed it back. She said, 'Don't be ridiculous. Take your time.'

She wondered whether we should all try to find supper in the old city. It was Ramadan, and since the city had fallen only two days earlier, most places were shut but there was a chance. She met me there a few hours later with some young Libyan activists. She persuaded a café to put a plastic table in the street, right next to the arch of Marcus Aurelius, bluffed her way past a Zintani militia group which had appropriated a courtyard house and, by taking great interest in the boss of an American television company, acquired us some extra food and Coca-Cola.

It was almost one in the morning when she wandered into Green Square. We followed. Her blonde hair was tied back in a ponytail, her sleeves were rolled up, revealing golden hairs on her brown arms. On her feet, under a pair of skinny jeans, were some soft slippers, and over her shoulder a backpack which contained her notebooks, water, camera and satellite phone.

People were firing heavy machine-guns into the air. A pick-up truck raced towards us. The radiator grille was missing, and there were bullet holes in the olive paint. Everything, including the windscreen, had been stripped off to give an unrestricted field of fire to the anti-aircraft gun pointing straight at us. The passenger in combat fatigues had a long black beard. They braked hard in front of us, and the man in fatigues leapt out. He grinned and greeted Marie. 'It's some of the Misrata boys,' she said. 'I think they've confused me with Portia.' Portia is twenty-seven and does not have an eye-patch.

When I heard this week that Marie had been killed in Aleppo, in Syria, I remembered her generosity, the lack of pomposity or competitiveness, the kindness she showed to younger journalists, to Libyan activists, even to travelling politicians. But what was strongest in both her character and her journalism was her ability to listen. When I returned to Tripoli this week, I found that my driver, the quiet, tall, elderly engineer in his battered car, whom the others had ignored, was now the prime minister. I'm sure Marie knew. But, still, I would have liked to have called and praised her.

GOVERNMENT LAND

As a member of parliament, chairman of a select committee and, for almost half my time, a government minister, I was often outside Cumbria. And as I moved 350 miles up and down on the train, between Westminster and the constituency, I oscillated between the question of how Cumbrians perceived politicians, and how politicians perceived Cumbria. I often felt as though I were translating between two incompatible languages: the language of policy (targets, metrics, frameworks, rollouts) and the language of place (this farm, that family, next winter's milk production). These following letters explore my evolving sense that the tools of modern government – which imagined clearly defined and understood economic sectors – struggled to operate in a place where half the economy was invisible to official statisticians, and where the most pressing local issue could be getting mobile coverage to a valley with thirty residents.

But I also became aware of more fundamental problems with our very idea of democracy. Some of the time British politics sounded very lofty: we spoke about 'justice' and 'equality', encouraging students to see politics in terms of the lives of Nelson Mandela or Gandhi. These men were often represented, in black-and-white photographs, next to Albert Einstein, on Cumbrian classroom walls. Most of the time, however, politics could not have been more different from that of these heroes in their struggles for social justice – a great deal

of energy, for example, was devoted to arguments over residents' parking. We lacked the rigour, the seriousness, or perhaps
the time to tease out the hope that lay between the melodramat-
ically grand and the mundane – to set an impressive and realistic
vision for what Cumbria could become.

Another thought. Aristotle talks about *logos* or reason in
politics, and this word suggests changing minds through facts
and arguments. In other words, he imagines a world in which
politicians would not be limited to mobilising their own base
and demonising their opponents, but instead try to persuade the
unpersuaded: to convince people of new ideas. To make them
open to a fresh perspective. This politics is mutual. The politician
should also be open to being persuaded or educated by the
citizen.

27 April 2013

Vote. For any party. Please vote. The turnout for our recent local elections has been in the 30 per cent range; the majority of the people reading this won't go to a polling station on 2 May.

I had a good afternoon on Saturday walking around Penrith. The sun was shining, and I must have knocked on about two hundred doors and spoken to about a hundred people. But among the excitement at a Sunderland goal, suggestions on the Penrith pong, or unexpected recommendations about Tunisian passport control procedures, I came across many who weren't voting.

I encountered anger, detachment and distraction. A man was refusing to vote in local county council elections because of rage about the Iraq war; a woman, because she felt that nobody had listened to her views on residents' parking. Some believed their vote made no difference; most that politicians made no difference. Those that felt they ought to vote seemed to feel it in the same way that they ought to go to church, visit their great-aunt, or take vitamin D; but they felt too busy.

This is, of course, disturbing. As generations were once taught, suffragettes died to win the right to vote. Everyone can intone 'government of the people, for the people, by the people'. But one man, one vote is less impressive when it is one man, no vote.

Indifference, however, is also understandable. Voters feel distant from their county councillors, distant from Westminster, very distant from a Member of the European Parliament in Brussels. It is difficult for most people to be sure what is the responsibility of the county council, the district council, Parliament, or the European Union. Government in different guises acts and fails to act, but why, and how? Policies are difficult to explain, more difficult to control. The representative is expected to represent the views of the individual voter, and also somehow the views of the whole constituency, and to balance their sense of the constituency's future and present against their views of their nation, their party and their own individual conscience. In a business or a regiment, unpopular decisions may be taken for granted; but in a democracy, voters want their own view expressed, and are inevitably often disappointed, sometimes by the decision, sometimes by the way it is made. In short, democracy never seems to be quite what people mean by democracy.

At the same time, people wonder about whether any of it really matters. Even at a national level, the grand fights have apparently been won. Catholics have been emancipated, women enfranchised, slaves freed. The fights for safety in the workplace, against feudalism and torture, are over. We are not at war at the moment. Most people are frustrated and irritated by a hundred things, but they do not rank them with the issues of 1832, 1911, 1940 or even 1983. And then there is the sense that politics, and in particular political parties, are a little shameful.

When I tried to invite some friends to help me deliver leaflets, I realised that they were looking at me in a new way. These people were my friends, but I felt that since I had become a politician, I had become something a little embarrassing, and not quite respectable. I received long, polite and very awkward replies, saying they were not sure how much they wanted to be associated with a politician or a party.

So why vote? The only answer, of course, can be that politics still matters. It is about more than a thousand technocratic decisions on where to spend, and where to save. Or about sounding old laments on deep, traditional drums. Politics matters, and will still matter, when the deficit is gone, and the NHS is fully funded, businesses flourishing, and GDP growth at 2½ per cent. Because politics at its best must always be about forging and reforging a civilisation. And, in our case, politics could be about forming a county, and a nation, that might still be extraordinary.

Few nations have ever possessed, on such a scale, such a proportion of educated, capable citizens. Almost every conversation on a sunny doorstep in spring reveals travel, learning or experience, rare even a generation ago. It is easy to talk about the hospice movement or mountain rescue; less easy to grasp the imagination that went into coming up with the ideas for these charities, or the determination and sometimes blind faith that keeps them still alive decades later. It is easy to record what has been achieved in community broadband projects, or affordable housing; it is more difficult to focus on how much research and flair went into making these realities. In a thousand ways, we prove, without quite realising it, just how much knowledge, power and skill we have, and could further develop, at the most local level.

Not just 'neighbourhood plans', but decades of education and experience have given us a potential no previous generation had, to define the future of this society at its most immediate, most local. Voting is not sufficient for any of this, but it is necessary for most of it. So at the risk of pomposity – however you vote, please vote. And vote locally.

My plea went unheeded – only 32 per cent of my constituents voted.

2 December 2017

When I lived outside Britain, I felt that the greatest problem in Britain was injustice. I could not understand how a country so rich could tolerate such poverty. Now I would add our shameless treatment of the homeless, the infirm elderly, and prisoners. But when I began to walk back and forth across the constituency – two months staying in private homes, visiting offices, holding public meetings; and then, two years later, set off on a 600-mile walk, recording three hours of conversations at a time – it was often difficult to reduce politics to big problems and big solutions.

When I was first elected, for example, I felt the key problem in our constituency was our remote location and sparse population: what I called the 'barriers of distance'. I felt that the solution was mobile and superfast broadband. I began by trying to fix things through the House of Commons: introducing parliamentary motions to oblige the government to provide mobile coverage; trying to make superfast broadband – like electricity or water – a basic right. Then, I tried to work as a fundraiser, to secure £40 million of investment for broadband in Cumbria. But I found that even £200 million would not be enough to deliver superfast broadband to every outlying farm. So I began to act more like a civil servant, focusing not on money but on implementation: looking at the costs of way-leaves, the way that EU 'state-aid regulations' prevented certain kinds of subsidy, the licensing of spectrum, the rights to the fibre beneath the rail network, the costs of point-to-point microwave links, and latency in satellite coverage. I went through a similar journey as flooding minister – from legislation on flood insurance, through securing extra flood money for Cumbria, to engaging with hundreds of experts.

As time went on, however, I began to wonder whether I would not have done better to get out of the way. Local communities, I decided, 'knew more, cared more, and could do more than distant officials'. We should involve them. So I pushed the Environment Agency to share more details of its flooding plans, encouraged the public to challenge them line by line, and set up dozens of consultation meetings to debate every scheme. Back in Parliament, I began to push for constitutional change – more deregulation and devolution to local areas, including giving them tax-raising powers and budgets.

I believed that by involving communities, everyone would ultimately form a single picture of the technical and financial challenges, and agree on the best solutions. But my community consultations did not seem to work that way. However much consulting we did, many intelligent well-informed people still disagreed very strongly with each other – and felt the government was letting them down.

I hoped later to discover through walking what 'the British people' wanted, but I did not find a thing called 'the British people'. Every household was defined by different experiences and very different notions. People were simultaneously happy and depressed, flourishing and frustrated. Many believed our problems could be summed up in words like 'poverty', 'austerity' or 'inequality'; that there was a simple set of villains – such as bankers – responsible for the mess; and that we could be saved by a hero with big ideas. But these concepts – justice or deregulation or devolution – rarely applied very neatly to a particular place or person or problem, even within the constituency. In fact, they seemed to mean contradictory things in different contexts.

Meanwhile, the demand for big projects and big ideas in Westminster remained insatiable. Every politician hoped to catch the imagination with 'big projects' – from the Millennium Dome to the nuclear facility at Hinkley Point – and with big ideas: 'Big Society', 'renationalisation'. Some of these ideas, however, were unworkable; few addressed the fundamental problems, and some meant nothing at all.

Practical solutions seemed instead to rely on a thousand detailed, complicated initiatives which never made it into a headline or a manifesto. While politicians produced phrases after the floods such as 'no more cuts', and 'this must never happen again', the real progress required reforms on hydrological modelling, weather reporting, water company reservoirs, dredging, soil absorption, tree-planting, bridge design, pumping, insurance, resilience measures, and the division of responsibilities between the Environment Agency and the county council.

I began to feel that political life – like our own families and jobs – is at its best a practical activity, not an ideology: a continual exercise of compassion, and grip, and competence, trying, as best you can, to do a dozen small things for each problem in turn. But politics seems to expect much bigger words and much grander promises.

25 December 2010

It has been a tough winter already and it is difficult not to think we are entering many years of harsh winters.

The snows cancelled some of the fairs, which I was looking forward to seeing, from Bampton to Brampton, and the Sunbeams group had to cancel its performance at Tirril. But, through all the cold and problems, it has still been a very musical Christmas, charged with different traditions.

A mother and daughter singing 'Once in Royal David's City' purely and beautifully in an ancient hall in Orton, with the father on the piano, could have been singing any time in the last hundred years. In Wigton, there were Bob Dylan solos in the Methodist chapel, a parade with a steel band, a rock group and a performance from *Joseph* from a Carlisle panto-mime star, before the carols began. In Penrith, the gentle light of the paper lanterns and the hundreds of voices from every corner of the square carried more ancient shadows and echoes.

The carol service in St Margaret's, Westminster, was also beautiful, but the end of Parliament was tumultuous. While I was making a speech in support of Cumbrian community pubs, a thousand students could be heard chanting outside, protesting against tuition fees, and missiles (which seemed to be snooker balls) were bouncing noisily off the roof of the chamber.

I enjoyed the *Any Questions?* radio show in Kendal. But two minutes before I went live on air, I slammed a finger in a heavy door, and my attention wavered a little between universities, criminal justice reform and guessing when my fingernail would drop off.

When I was asked what book I would give a child for Christmas, my mind went blank. I finally suggested a book about the young King Arthur that I enjoyed as a seven-year-old. But the more experienced politicians who followed me apparently had much more advanced ideas for modern children.

One recommended *Schindler's List* and Andy Burnham suggested the ideal stocking would include the socialist parable *The Ragged Trousered Philanthropists* and Conrad's account of the African slave trade, *Heart of Darkness*. Since Andy Burnham had been Health Secretary, I showed him my finger off-stage and he replied: 'I was a spin-doctor, I'm afraid, not a doctor.' When I showed the finger to Catherine in my office, she cried: 'You've done it on purpose, haven't you? To get out of signing the Christmas cards . . .'

Penrith community hospital did a wonderful repair job on the Saturday morning, and we have been able to send out a Christmas card, designed by Ellie Morton, who is six and from Plumpton School. The angels are set against a dark blue sky.

My favourite event was the fair at Brough. More than five hundred out of a total Brough population of seven hundred are involved in putting on the fair. There were men wobbling on penny farthings, a youth committee raising money for a kick-wall, children in Victorian fancy dress, a long tail competition for dogs and a church room with sticky toddlers making Christmas decorations.

The parade featured vintage tractors, a large steel band (led by a man in orange with an electric guitar and a cowboy hat), a very smart detachment of air cadets with their colours (one of them has just been selected for pilot training), dancing girls in leopard-skin hoods, mobility scooters, the scouts and a Shetland pony pulling Santa. A vintage bus ran services back and forth to Kirkby Stephen.

The Archdeacon of Carlisle, the chairman of the district council and the mayor led the procession. The fair was opened by the High Sheriff of Cumbria, in knickerbockers and sword, reading the medieval charter of Brough, issued by King Edward III in 1352. And in case anyone thought it was too serious, Queen Victoria appeared in full black bombazine to take the salute.

Nor will I forget hobbling through my hall on the Sunday morning of the Penrith Christmas lights switch-on, in stiff orange plastic boots, carrying my skis. The sky was such a dark blue, the sun so bright and the snow so crisp that I felt I was at 6,000 feet. Only the great ash trees, the low white-washed farmhouses and the trembling Swaledale ewes reminded me that I was outside my own front door. Soon, instead of splashing through soft, uneven ground, I was gliding over snow.

The hoof-tracks of fell ponies and the bleached moss on the boulders were all hidden. The sunlight gleamed off the ice-crystals. And with each plant of my ski pole, each thrust from one ski to another, each breath, the world grew simpler and more silent. When, on Loadpot Hill, I stripped off the skins and skied down again, I could see russet tips of grass poking through the snow-crust. I half-recognised limestone ridges as

I skied over them, and I had to hop flowing becks. Close to home, I lifted my head and absorbed the changing shapes of the white valley fields as they rushed towards me. The dark blue sky was the same colour that Ellie had painted behind her angels.

21 June 2014

Last Saturday, Shoshana and I spent the day at the Cumberland show. It has passed its 175th anniversary and you could see why it has lasted so long. It wasn't only the ice cream. We must have met two hundred people who had come from across 2,000 square miles of Cumbria. Shoshana stood, transfixed by the sheepdogs who were herding geese, and the gundogs who were finding sausages in piles of branches. She seemed to admire every muscle of the Charolais, and to photograph many hairy feet of many Clydesdales. I don't seem to be able to keep her away from shows: last year she came with me, I think, to Dufton, Lowther, Penrith, Skelton and Hesket Newmarket. Now she is bringing her mother all the way from the United States to see the Skelton show again on 5 July.

But it's also great to see how much she, who was a teacher in inner-city schools in the United States, seems to be impressed by British schools. The day before the Cumberland show, we were both at William Howard, in Brampton. Last time I was there, I was shown a mobile science lab with state-of-the-art optical equipment. The time before, I was in the dance and theatre class. This time we attended a class in financial management; heard a twelve-year-old debating interest rates; saw a video on financial education, shot and edited by another member of the class; and ate cakes which the children had baked for elderly people from Brampton. (They had invited them all to the school for a mass tea party.)

Schools are, apart from the Church, the most long-lasting institutions in our society: with deeper textures even than the agricultural shows. So, although Greystoke primary school is celebrating only its 175th anniversary – it was founded like the Cumberland show in the *annus mirabilis* of 1859 – it celebrated with members of the Howard family that had founded the school, who still live in the village and whose ancestors on the maternal line were already the Lords of Greystoke at the time of the Norman conquest in 1066. On the way in to watch the Appleby school musical, I passed a copy of its 22 March 1574 charter on the wall (its origins go back to the thirteenth century). At Nelson Thomlinson, I saw copperplate records from 1914 recording teachers stuck in snow and students sent to the First World War, next to a classroom containing student entrepreneurs who had just built a software training program, which they ran from a computer the size of a playing card.

At Fellview primary, in Caldbeck, ten-year-olds asked me about dignity and trust in politics. At Penruddock, I was cross-questioned on the constitution. In Armathwaite, I spent forty minutes with a class, discussing philosophy, peacekeeping and the causes of war. Shoshana made me appreciate again how lovely the settings were. Yanwath's new playground lies in the gently rolling land where the Lowther approaches the Eamont; Milburn's stone building sits, square as a Norman tower, in the very centre of the village green. And Armathwaite's broad, wood-fringed playing field is perfectly framed by the fells.

Then I returned to Westminster to chair the Defence Committee. We welcomed 7th Brigade back from Afghanistan, hosted Gurkha cadets (these eighteen-year-olds from remote rural areas of mountainous Nepal asked me in fluent English about the European economic crisis, and our policy towards Syria); saw the defence attachés from France, Australia, Germany, the Netherlands, Israel, Canada and Denmark; cross-questioned the two outgoing chiefs of the Defence Staff; and held a debate on the floor of the House, pushing for higher defence spending, more focus on the dangerous implications of Russia's annexation of Crimea, and the ISIS takeover of western Iraq.

Would the children at Armathwaite primary school sympathise with this push for more defence spending? They seemed, on that sunny morning above the loveliest stretch of the Eden, very far from any kind of war. Child after child questioned whether violence could ever be the answer. I found it difficult to explain what I and my colleagues on the defence committee took for granted: that we were now facing an extreme threat from Russia that required us to spend far more. And I wonder how we will know which of us is right.

But perhaps the most striking thing last week was the fiftieth anniversary of the Lazonby swimming pool. It, too, was the product of a primary school: the project was driven by the local head, John Hume, in 1964. It had been built by the community, for about a quarter of the price quoted by a contractor, because the community provided the labour. It, too, had changed with the modern world: the green river water had been replaced with mains water, fences had gone up, a playground had been built, health and safety now required a full-time lifeguard.

The new generation of volunteers who were triumphantly keeping the pool open were now facing regulations and restrictions which didn't exist before. But Mr Hume who had built the pool, fifty years ago, was still there. All that effort and imagination was still preserved; Mr Hume, at ninety-two, swam a lap with the children.

1 April 2017

About every three weeks, I seem to be perched on the edge of a desk – sometimes a comically small desk – talking politics in a classroom. Some Cumbrian schools have only twelve students, some have 1,500. But the conversations are strikingly similar.

Eight-year-olds begin with broadband; sixteen-year-olds with reducing the voting age to sixteen. And then – quite suddenly – the conversation becomes global.

In a moment, I can be asked to sign a pledge to eliminate child poverty in Africa, or guarantee 100 per cent access for girls' education in Asia. So, the conversation lurches from the very local to the excessively international without touching anything in between.

There is no mention of the deficit or the budget. No mention of election campaigns, political parties, the civil service, local councils, the law, or – still less – politicians. No mention, in short, of how Britain is actually governed.

And I realise, thinking about it now, that it isn't only school-children who have very little understanding of our political system; I myself knew almost nothing until I became a politician. And when I actually had to sit through the speeches, see how the voting happened, how promotions worked or 'whipping' operated, I was often bewildered, disappointed or even horrified.

Week after week I was forced to see how powerful solutions from intelligent, hard-working people were defeated: by chance or timing, by an unfair headline or a change in prime ministerial priorities, by the unintended consequences of regulations, or the bizarre logic of the law and the inertia of bureaucracies.

How should you teach politics or citizenship in school? Some Cumbrian teachers seem to favour classes on human rights, others inspiring stories of political change-makers. Some encourage children to campaign, to sign petitions, to go on demonstrations or mount social media storms – focusing on getting their voices heard. There are many black-and-white photographs of heroic imprisoned freedom fighters such as Gandhi and Mandela. But something is missing. If the aim is to create citizens who can really change the world, then I feel we should begin by helping them understand the grinding reality of democratic politics in modern Britain.

So, my starting point for a 'citizenship' curriculum would be a few episodes of the imperishable 1980s government satire, *Yes Minister*. My first exercise would be for the pupils to name their best political idea, and then write a tabloid story ripping it to shreds. Then I would invite them to go through the week of a typical politician, beginning with voting at 10 p.m. on a Monday in Westminster, through committee hearings, debates or ministerial business cases, to a formal dinner three hundred miles away in the constituency on a Friday night, inserting a little time for the family.

I would make them watch a *Question Time* politician giving answers on cancer survival rates, local government financing and defence policy in NATO, and explain the bluff behind it – make it clear how little a politician can ever be expected to really know. I would invite them to draft replies for just one day of my email inbox. My predecessor Willie Whitelaw received perhaps a hundred letters from constituents in a year. I am not unusual as a modern MP in receiving and replying to perhaps 20,000 emails from constituents in a typical year. And then I would suggest watching *Yes Minister* again.

But that is the easy bit; two difficult tasks would remain. The first is to explain how British politics can be somehow more than the sum of these dispiriting parts. There was no golden age of political dignity and respect. Eighteenth-century visitors to Britain were appalled by the rudeness, the cynicism and the irresponsibility of the British press; they were scandalised by the brash self-assertion of the party manifestos; and revolted by the boisterousness and lack of respect in the chamber. But these notorious frailties somehow helped to keep our system more alert and responsive than was possible within the grandeur of the American constitution, or the logic of the French.

The second task would be to help the student see how – despite all the apparently unmovable forces – real political change remains possible. I cannot describe how the whole thing, which we call our 'constitution', manages to survive. But if I was trying to write a 'citizenship' course for students I would start with that – with the infuriating, muddled eccentricity of parliamentary procedure and parties and media and civil servants and half-informed ministers, through which wars have been won, peace concluded, and treaties made and broken; through which a National Health Service was created, and slavery ended.

28 May 2011

I have been in Parliament for over a year but I am only just beginning to understand how it works. It speaks: all the time. It is a last
fragment of pre-literate England: a place where what matters is
not what is written, still less what is emailed, but what is spoken.
Speaking is the currency of the House. But who is listening?

According to a website, I have made eighty-five 'appear-
ances' in Parliament in the last year, and spoken on average in
one debate a week when Parliament is sitting. I have spoken in
topical and oral questions; in Prime Minister's Questions and
urgent statements; finance bills, adjournment, opposition and
Westminster Hall debates; delegated legislation and European
committees; and spent more than a hundred hours on a select
committee.

Sometimes, I'm struggling to take nineteen interventions
in a fifteen-minute speech beneath the wood panels of Westminster Hall; sometimes I am cross-questioning a minister in a
room with pop-art tapestry on the wall. Sometimes I am asking
the prime minister a question at midday on live TV and the
chamber packed to standing; sometimes it is 11 at night and
there are only the six Cumbrian MPs present.

I have spoken on single farm payments and flooding; on
localism and Libya; pubs and Afghanistan; the budget and Brazil;

on funding for charities and for special educational needs; on the Olympics and shootings and broadband. I have learned whether to call the person at the end of the room Mr Speaker, Chairman or Ms Primarolo; and that, when the mace is under the table, I should speak only to 'the amendment'. A website has concluded that I have used three-word alliterative phrases ('she sells sea-shells') twenty-six times. Even my mother is beginning to get tired of watching me on BBC Parliament, but many of my colleagues speak much more. What, however, is the point of all this talking?

Last week, I introduced my own debate on mobile phone coverage on the floor of the House. At present, at any one time, more than 7 million people cannot connect to a mobile signal. The problem is at its worst in rural areas. Good mobile and broadband coverage will help our thousands of tiny companies. Mobile signals can allow us to use devices to monitor our health, minute-by-minute around our home, instead of being trapped in hospitals. Online learning can transform children's performance in education.

For Cumbria, mobile broadband will improve economic growth and our overstretched services, and help us live and flourish in remote areas. But Ofcom, the mobile regulator, has no plans to extend the coverage. Why? Because the mobile companies complain that it is expensive maintaining masts in rural areas where they have fewer users, and the companies threaten that they will pay 5 or 8 per cent less for the licence permissions if the government forces them to cover rural areas. This would mean that the Treasury could receive only £2 billion for the sale instead of £2.1 billion. So, for the sake of one or two hundred million pounds, the Treasury proposes to leave millions of people and thousands of square miles of Britain without coverage, probably for the next generation.

At first, I tried to simply convince Ofcom to change its mind through private meetings. I called a dozen experts, catching them in China or from the Eurostar, and talked to them until my mobile burned against my ear. I pored over technical papers on how to split, stretch and retransmit the spectrum. I met with civil servants, with junior and senior officials at Ofcom, with ministers and with telecoms companies. At first everything seemed fine. But then senior Members of Parliament began to say, quietly, 'You are not going to make any friends if you keep pushing.'

I couldn't believe it. In return for making slightly less in the auction, we could have a mobile broadband infrastructure which would bring great social and economic benefit. Surely, it was a no-brainer. But I was soon hearing a dozen arguments against: that the spectrum was 'unsuitable for voice'; that an increased obligation would cripple the companies and 'undermine urban coverage'; that it could be easily done in three years' time instead. Every argument was untrue. But it was clear that I was getting nowhere.

So, I pushed for a full debate in the House of Commons. I spent days bringing it together, writing individually to 150 MPs, convincing the House of Commons to create time, negotiating with the department and the whips and organising speakers. Last Thursday, I opened and closed the debate; 110 MPs from all parties signed: more than on any motion debated on the floor of the House in living memory; thirty of these MPs spoke. And we passed the motion to urge Ofcom 'to increase the coverage above 98 per cent of the population' with a unanimous vote.

Here, it seemed, was a real concrete use of Parliament. But yesterday, a colleague said nothing would happen for fifteen years. The head of Ofcom is angry with me; officials are referring mysteriously to complications with 'coverage definition' and 'the distortion of market diversity'. And the whole thing seems to be slipping out of my hands again. I am back to writing letters and emails and lobbying and pleading with officials and ministers. I think I'll win this fight, eventually, but Parliament has spoken. Who, really, is listening?

I'm still not sure whether this vote in the House of Commons had any impact at all. But something changed. In 2013, partly in response to our parliamentary campaign, Ofcom demanded in the 4G spectrum auction that one of the bidders covered 98 per cent of the UK population.

This still excluded one and a half million people, and thousands of square miles of the UK, much of it in Cumbria. The mobile operators continued to complain about the costs of installing fibre and masts. That year, therefore, I convinced George Osborne to allocate £200 million for building rural mobile masts. He agreed, but the project was cancelled when he stepped down – leaving Bewcastle with the foundations of a mast but no signal. My attempts to force the mobile networks to share their fibre and masts in rural areas were also unsuccessful.

The year after I stepped down, however, the networks finally agreed to share services. For the first time targets were agreed on land mass, not simply population. As of now, about 95 per cent of the land mass has been covered, and allegedly 99 per cent of the population. I suspect that many ministers, councillors, civil servants and regulators will feel that they were responsible for this change. And if they were even conscious of our complicated campaign and parliamentary vote, they would feel it had been irrelevant.

27 September 2016

My father liked to say, to enthusiastic nods, 'We don't live in a democracy, you know.' He meant, I think, that politicians were not representing the public accurately or effectively. Why? Because politicians were some combination of – to use some of his favourite phrases – 'out of touch, ignorant, idle, self-interested and "wet"'.

Modern British politics was the opposite of everything he believed in. He couldn't see the point of an MP going to Parliament, which he thought was a 'giant talking shop'. Nor of passing laws ('there are too many already'). He saw the 'separation of powers' – between central government, county and district government, and between ministers, civil servants and police, as 'a recipe for disaster', meaning consultations and committees, with everyone vaguely talking round issues, passing the buck and getting nothing done. He felt that most of our laws did not protect the public, but instead led to endless delays, providing fees for lawyers and excuses for idle civil servants. He felt that 'civil society' – the media and NGOs – were not a great feature of democracy but instead a general contribution to the atmosphere of negativity, 'back-seat driving' and 'second-guessing'. He thought that all the checks and balances and constitutional protections – put in place to frustrate absolute power, or tame the Trumps – essentially prevented the public getting what they wanted because they prevented anything getting done.

His ideal was instead for a Member of Parliament to directly govern a patch of territory: be out and about every day meeting locals, hearing their complaints, and vigorously sorting them out. In his ideal constitution, the community would be small enough for the MP to be able to meet a good percentage of his voters. The job would not be interrupted by duties in Parliament or a ministry, so the MP would be in the constituency full-time. Instead of relying on government civil servants, he would recruit a band of volunteers to help him – ranging from Boy Scouts to energetic retired people. Together they could provide practical, rapid solutions to the public's problems. If there was a problem with crime, he would go out with the police to catch the criminals. If there was a shortage of schools or affordable housing, he would build some – immediately, without too much fuss, and possibly with cheap asbestos roofs.

My father called this a 'real democracy' but it didn't have much room for laws, bureaucratic regulations, civil servants, or even votes. In fact, it looked suspiciously like the job he had done as a district commissioner in colonial Malaya – where, he told me, he had ignored procedures and laws, done what he thought was best, and trusted that if he succeeded, no one would ask too many questions about his methods.

My anxiety about democracy was different. I thought the fundamental problem was not that government was slow but that it was unrepresentative. Democracy apparently meant representing the people, but which people? In theory, the MPs represent all their constituency to Parliament, and, in government, represent all of the British people. But what do 'all' Cumbrians, let alone all the 'British people', actually think or want? And how would you find out? Not through the two hundred letters an MP might receive in a week, when 60,000 constituents haven't written in. And although opinion polls are theoretically more 'objective' than a post-bag, we have seen three times recently (with Scotland, the election and Brexit) how wrong they can be in predicting public opinion. At best, they give only very simplified summaries of people's views – and in any case, on almost any important issue, opinion seems divided pretty equally.

Which is why, my father might reply, a politician 'should ignore public opinion, and simply do what they think best, by showing leadership'. But democratic politics is not an exercise in science or engineering – there is no simple 'best solution' for people, independent of the views of the people themselves. A doctor might think community hospitals are an inefficient allocation of resources, but the public has every right to feel that for a dozen reasons, they still want to retain beds in their local community hospital. An economist might conclude that it was not cost-efficient to provide superfast broadband to remote communities (at £40,000 a house), but the public might feel that broadband is a basic utility and right, like water and electricity, and be prepared to bear the cost. A climate scientist might favour wind turbines, a community might prefer to generate energy by other renewable means. And being a democratic politician means listening as generously and imaginatively as possible to the public. Not – as my father often implied – simply to get re-elected, but because 'listening' is in fact the job description. It is what makes being a politician quite different from being a 'governance expert'.

Above all, it seems to me, politics is about trying to work alongside the public to shape and define what kind of society we wish our children to inhabit. Normally that conversation is reduced to phrases like 'creating jobs', 'eliminating poverty', 'creating a more equal society' or 'growing the economy'. But of course, that only catches a tiny percentage of the details of what we actually want in our personal lives. A better local park, yes, but also beauty, or opportunities to be frightened, inspired and entranced. For many of us, perhaps, even a chance to show honour and courage. Nor does it capture what we might want not for ourselves as individuals but for our nation – our sense of duty, of international consequence, or positive pride. Or the way in which these in turn draw on facts of our history, features of our soil, shards of our national myths, and our own experiences at home and abroad – a mixture muddled in all of us, and fiercely disputed by people with different tastes and beliefs, and from different generations.

Pace my father, Members of Parliament are not at their best when they are being a decisive district commissioner or even a CEO, but when they are engaged in a joint project with citizens of creating a story of the nation's future – which is never entirely objective, but is better or worse depending on how truthful it is to real experience, how complex, how modest, how flexible, how capable of penetrating from the superficial opinion to the deeper, more serious instincts of us all.

8 December 2012

On Sunday I sat with six professors who were discussing how 'to change the world'. They included a Central European dissident, a computer specialist and a seventy-five-year-old French communist. At times they seemed hardly conscious of each other, and the words they used were puzzling (the Frenchman, for example, liked to talk about 'the happiness of dissatisfaction'). But it was also moving.

At least they were trying to think about changing the world. For most of us, perhaps, most of the time, the world is too complicated. I don't understand how a microscopic speck of liquid can transform in less than nine months into something, like us, with rigid bones, hair and fingernails. I don't understand what an element is, let alone quivering quarks, which have a place without time, and a time without place. And that's before you get on to Norway's European policy, or Chinese trade. Or whether we have free will. The overwhelming complexity of the world may tempt us to ignore the bigger picture and just focus on our particular job and our own particular family.

But these professors knew a lot, spoke well about philosophy, psychology and history; and they had ideas about human nature, and had the confidence to debate world change, and not in the comforting slogans of a self-help book.

All agreed that something very important was missing in our modern lives. For the French philosopher, the problem was that we were too obedient, that our democracies were simply a theatre, behind which all the decisions were taken by a small elite, and that we had forgotten how to be happy. The American said that the problem was that we were atheists who had lost a sense of the sacred. The psychologist said that we had become manic, and lost our sense of balance. The computer specialist said that we treated our society as a 'bug' which could be fixed; and we had lost humility.

Everyone agreed that we have become lonelier and more obsessed with money. The Frenchman was the only person with a clear solution. He said we will only become happy through revolution and the pursuit of communism.

I suspect, however, that something is missing in this discussion of ideology and global change. They were correct that the modern world makes individuals more isolated, and that many of us have retreated to a space that does not stretch much beyond the office or the front door. And that we often feel powerless. But they are wrong in implying that this is because our power has been removed and given to someone else. My sense, instead, is that the problem with the modern world is that there is no power anywhere. In my experience, politicians, journalists, civil servants, farmers, bishops, and even – difficult though it is to believe – bankers often feel powerless. The 'elites' believe that they are trapped in an immovable – if slightly wobbly – jelly of conventions, regulations and procedures. Change feels almost impossible. Many of us feel increasingly inert.

But Cumbrians seem to be the exception. We are surrounded by Cumbrians who are taking responsibility for projects which in other parts of the country would be left to state specialists.

We have parishes challenging central decisions on cycle paths, on winter grittings, and if they don't like the answer, doing it themselves. We also have some of the very first villages in the country buying and running their own cooperative pubs and breweries, setting up their own independent planning policy and designing their own fibre-optic cable networks. Cumbrians are moving into specialist areas of education (where we have a record number of schools going independent of the county council), housing (some of the country's first community land trusts) and health (we have the only community-owned ambulance in the country).

We often perceive this as something forced on us, just a way of getting a job done. But we underestimate how much the organising process to change an area changes us as people.

The Frenchman may be right. We may be happiest as humans when we are dissatisfied, want something to change, and become personally involved in making that change. It is not only that working alongside your neighbours is less lonely; it is that driving a project for which you are not paid, and which benefits other people, brings dignity. And building something in your neighbourhood – a hall, a school, an affordable housing estate or even a broadband network – involves you in a sequence of practical challenges, results in something visible and tangible: a communal achievement you can see and live alongside. It brings a positive sense of pride.

But could we extend this vision to the national and international stage? Could local action become a bridge into participating in much larger struggles and choices? Could we move, for example, from simply supporting a local school, community hospital or housing scheme, into shaping national policies on education, health care and architecture? Could the local be the best place from which to do such things? To understand not only what dissatisfies us about the modern world, but also decide what we would like the world to be. Could the way to change the world be to begin in Cumbria?

Edinburgh, 3 February 2018

Last Saturday was Burns Night – trews, kilts, tartan sashes, a haggis, an ode, two pipers struggling to finish the whisky in their quaichs. Three hundred heads nodded wisely as someone sang 'Robin Shure in hairst' and perhaps ten heads knew what those words meant. Why were we ritually remembering Rabbie Burns?

After all, the poet would have seemed to his contemporaries not so much ahead of his time, as centuries behind it. He had never seen a city when he first stepped foot in Edinburgh, aged twenty-seven, in 1786. The world around him was in motion – Scotland was changing from the poorest country in Europe to almost the richest; everyone was on the move. Fifty per cent of the people of central Scotland would soon be living in places where they had not been born; an 'epidemic' of emigration was already driving Scots to America.

But Burns had remained resolutely fixed in rural Ayrshire, ploughing his father's seven-acre farm. (He left school at twelve.) Small farming was, for all his reputation with the ladies, his most enduring and dangerous love. He dressed like a farmer, in rough clothes and boots, with an unruly mop of dark hair. And he spoke and wrote in an Ayrshire dialect, sprinkled with words which had barely been written down since the sixteenth century, if at all. He was entering a city which was the Silicon Valley of its day. Edinburgh had the best university on the planet. It was where you went if you wanted to study at the greatest medical school, or work with the men who were reinventing religion and inventing economics.

The celebrities of Edinburgh had left the dark, dangerous, narrow streets of the old town on the castle hill, and had built a utopia out of open fields – square upon square of the very latest and most opulent architecture, filled with light and air – and called it simply 'the New Town'. The land for the shops, selling the latest French fashion, went at £120,000 an acre.

The New Town seemed to reject every aspect of Burns: from his old-fashioned provincial dialect to his rough clothes, medieval verse forms and bawdy humour. Edinburgh's intellectuals saw themselves as citizens of the modern world. And it was their ideas from their universities which they expected to define the century ahead.

Burns, who cared little for much of this and retained an embarrassing romantic loyalty to the drunken, exiled pretender Bonnie Prince Charlie, reminded them of a past that they liked to forget. Little wonder that Burns did not stay in Edinburgh, nor take the offer of going to London. And although his poetry hums with the rhythms that defined the French revolution – 'a man's a man for a' that' – he never visited Paris. Instead, he took his book royalties and invested them in a few more acres of sour upland soil, which he ploughed with his brother.

The Edinburgh professors could have explained to him through their economic textbooks and treatises on the agricultural revolution the reason why Burns's small farm was doomed to failure, and why he was eventually forced to sell and take a job as a minor civil servant in Dumfries. Their medical textbooks might also have correctly predicted that he would die young (with too many children, by too many women), sick and deeply in debt.

And yet it was Burns's style, from his simple clothes, open-necked shirt and natural hair to his plain assertion of love – romantic love – which defined the next two centuries. It was his 'radical' views on slavery and women's rights that became the basic assumptions of our age. And it was his simple background, his background, his informality, his celebration of sex, his rough language, his celebrity and his early death which made him the earliest image of a rock star.

How quickly the tortuous hundred-word sentences of the Edinburgh elite became unreadable. How rapidly their plum-coloured knee-britches, their white silk stockings and vast buckled shoes, their peacock-blue morning coats and Chinese brocade waistcoats became outdated. Within a decade their entire culture seemed the relic of a surreal dream: a dream in which a white powdered wig on a shaven pate could seem the most attractive way of treating a young man's head. (When Burns put on a hat he looked like a cowboy.)

Burns is not so much a fading echo of 'Auld Lang Syne', but a warning to every ancient Athens and contemporary Silicon Valley. A warning that just as you have convinced yourself that your concerns – Platonic forms or robotics, rhetoric or AI, your diet, your dress, your companies – will take over the world, you are on the precipice of history. While the very people you imagine have failed to break free from the past have, in fact, defined the future.

BORDERLAND

THESE LETTERS FOCUS ON THE English–Scottish referendum. I am half-English and half-Scottish, with a family house in Scotland. I represented the only constituency with 'border' in its name. I served very briefly in a Highland regiment, before going to an English university. Keeping the United Kingdom together felt to me like the most important political campaign of my career – even more important than Brexit.

Penrith and the Border, as I keep repeating, was historically the heart of a Middleland nation which once stretched from my English constituency at times almost to Glasgow, and which had spoken a language completely different from Gaelic, Scots or English – and had its own epics, kings and saints that had nothing to do with England or Scotland. This sense of a shared identity on either side of the modern frontier, centred on the Debatable Lands, had persisted right through the wars between England and Scotland, in the Border law and the culture of the reivers. Faced with a culture which could not be divided on conventional lines, the medieval Scottish kings and lawyers had simply called the Cumbrians 'the Britons'.

It was, therefore, a place where I could feel British. I had hoped through my walking to unearth a rich, deep and distinct cross-border identity. But what I found was often a shared amnesia – and an almost entirely virtual nationalism – which evaded any definite stories about blood, history or soil, so ambiguous and indistinct as to be hardly an identity at all, beyond a rejection of some hated – but equally elusive – other.

1 March 2014

A trainee teacher: 'Scotland separating from the United Kingdom? So? Does it matter?'

A retired general: 'I'm not saying I believe this, but there are some people who think it might be quite good for England. That Scotland is taking advantage of us.'

An artist: 'I'm not that political. Shouldn't you be asking a politician?'

A newspaper editor: 'We haven't decided our editorial position. We won't decide whether the newspaper will be backing the United Kingdom until much closer to the vote.'

The head of a large membership organisation: 'I'm afraid we can't be seen to be "political". It's complicated.'

But on 19 September we could wake up and find that Scotland has separated. That the United Kingdom as we have known it for four hundred years has vanished. That a third of the land mass of Britain has broken apart. Officials will be racing north and south to negotiate the terms of the separation: working out what to do with the nuclear submarines in Scotland, with the national electric grid, with currency and passports.

Anyone with mixed English or Scottish blood will have to choose a single new identity, and reject a part of their previous identity. I and a million other Scots in England will have to decide either to return 'home', to try to engage in the very difficult and uncertain task of launching a new, much smaller country; or to make a new life in England as an immigrant in a foreign land.

Why, given all this, has there not yet been a mass demonstration of support for the Union? Our population has never been so educated, long-lived, confident, well-travelled or well-informed. We are concerned with poverty in Africa, nuclear waste, the environment, the welfare state and superfast broadband. A change to the lobbying bill, or a threat to the public forest estate, can fill an MP's inbox with thousands of emails. A proposal to build a wind turbine can bring a hundred people in an instant onto a windy moor in the rain. A million people demonstrated against the Iraq war; several hundreds of thousands demonstrated against the hunting ban. Voters are rarely shy to say what their values are, or what they want for the United Kingdom. So why is there so little energy in saving the United Kingdom itself?

When states tried to secede from the United States, the reaction was civil war. Even in gentle, understated Canada, hundreds of thousands of Canadians from outside Quebec rallied before the referendum to plead with Quebec to stay. Spain or India would not countenance Catalonia or Kashmir breaking away. You might expect every major British public figure – editor, writer, filmmaker, actor, scientist, sportsperson, historian – to be making their own passionate and sincere arguments for why they care about Britain. But it just isn't happening.

It is tempting to blame apathy; but I suspect the problem is partly our identity. In most of Europe, nationalists worked in the nineteenth and twentieth centuries to simplify their identities: invented new governments and reintroduced old languages, moved borders and then cleared populations to eliminate diversity. Mixed territories – the Austro-Hungarian Empire or much more recently Yugoslavia – were broken into smaller 'nation states'. The idea was to ensure that someone who might once have been simultaneously German, Czech, Czecho-Slovak and Austro-Hungarian became 'simply' Czech.

But here it was different. The United Kingdom continues to incorporate four formal nations, four languages, and other half-forgotten kingdoms that lie beneath those later nations, of which Cornwall and Cumbria are two. Our queens and kings have ancestors from all those nations, and from dozens of European nations too. This idea of a United Kingdom uniting different nations is the deep grammar behind our lives, our dreams and our actions. It is the context of our democracy, and the framework of all our government. Soldiers sign up to serve and if necessary die for the Crown – which is the Crown of the United Kingdom. When we vote in a general election, we are choosing people to represent the United Kingdom.

The United Kingdom is precisely what we have for more than three hundred years been working to improve and preserve. After so many centuries it is very difficult to imagine what our identity would be without it. But none of this makes the United Kingdom any easier to understand.

We feel proudly English, or Scottish and also British, in different bewildering combinations. We have the same national broadcaster, the BBC, but separate legal systems. We compete against each other in rugby but alongside each other in the Olympics. Cumbrians sometimes talk about Britain, sometimes about the United Kingdom, sometimes about England. We have forged an identity which is contradictory, complicated – including both the Shetlands and London, where 60 per cent of people were not born in the United Kingdom.

This complexity is not something to try to deny, or 'simplify'. It is something we should embrace. Not through pompous pieties or faking a rainbow community. We need different voices – poets, musicians, sportspeople and community groups, from both sides of the border: people who, unlike politicians, are capable of being passionate about our predicament.

The different nations embedded in this Kingdom should be able to be outrageous, even rude to each other, without erasing our future together. We should revel in the complications, the oddities of our borders and the contradictory names we give our islands. We can hate each other's sports teams, without hating each other. We – England, Wales, Scotland and Northern Ireland and older, more interesting places like Cumbria – experience this tension, muddle, discord and love, precisely because we are a family. We should not abandon that 400-year-old relationship, just because it's not 'simple'. We have never been simple people. And perhaps that is the strength of the United Kingdom, in a world that isn't simple either.

25 May 2013

Recently, I forded the Solway from near the King Edward I
memorial to examine the border between England and Scotland.
I stopped, two hundred yards beyond the shore, in salt water up
to my waist. The tide had gone out, and the distance to Scotland
was half a mile of sea. What would I find on the other shore?
Eighteen hundred years ago, the beach off which I had stepped
was Rome, the shore up which I was about to climb was Barbarian. You would have stepped into the water from a place with
Roman senators, legions, taxes, magistrates, villas and temples,
and have emerged into a place with no such things. One land,
one culture, one nation and one state stopped at one shore, and
at the other a completely different set of institutions and powers
began.

But walking out of the sea into Scotland 1,800 years later,
I saw no one. The sea chopped against a slag-heap of uneven
boulders, glistening black, algae-covered, just below the scum of
bird feathers that marked the high tide. On the joins between the
coast and the fields were thickets of briars and nettles, wild rasp-
berry bushes and hints of more exotic, pink-flowered aliens: the
same plants you can see in a thousand railway sidings, builders'
yards, landfill sites and canals.

After an hour, I met an eighty-year-old Scottish farmer. He

was polite but he did not have much to say about the border, or the difference between the two countries. And after five hours' walk, I crossed the border at Gretna, re-entering England, confused.

After the collapse of Rome, the differences between the two sides of the Solway became less stark. The little state of Cumbria stretched from Eamont Bridge sometimes almost to Glasgow, and Northumbria from Newcastle to Edinburgh. Right up to the Highland line, both sides came, by the thirteenth century, to speak the same language – English – and wear the same clothes. A single economy was developed by European monasteries with abbeys on both sides of the border.

The frontier at the River Sark was the product of the thirteenth-century wars between Edward I of England and the claimants to the Scottish throne, John Balliol and Robert the Bruce – the moment when, after almost a thousand years of a cross-border culture, something like the Roman frontier was erected again. The ambiguous claims of England to Scotland and Scotland to Cumbria fell away.

On one side, the Scottish state possessed absolute power, but at the millimetre line of the border, its sovereignty ended. On this side of the line, the English were citizens in their own nation – it was their home, from which nobody had the right to drive them, and their government had obligations towards them. Step one foot across the mid-point of the Solway, and they were aliens. They were no longer citizens under their own parliament, judged by their own judges under their own laws. They could no longer choose their representatives and stand for office. In a single step, they were under the jurisdiction of a different law and a different government. At worst, they were enemies; at best, guests and strangers.

The two nations were reconnected in 1604 when the Scottish

king, descended from Robert the Bruce and his enemy, King
Edward I, became king of England as well. He, James VI and I,
is the ancestor of Queen Elizabeth II and her son Prince Charles.
A century later, we became a United Kingdom, with a single
parliament in Westminster. There was no longer a need for a
border. No fence. No border police. Because we were Britons –
citizens equally on both sides of the border, with a sense of
their rights, in their own country. King James called our area the
Middle Shires. In total, we have been separated by the Romans
for three hundred years, reconnected in a Middleland for almost
a millennium, then separated for a few hundred years, and then
connected again with a single head of state for the last four
hundred years. Separations have been relatively brief interludes –
a very long time ago.

So why did things still seem to change so starkly when I
reached people on the other side of the Solway marsh? Like
us, they shopped at Tesco's, spoke English, watched the BBC
and grumbled about Westminster. But the people I met on the
far shore spoke with a different – Scottish – accent, had been
educated to take Scottish exams. They supported a separate
football league. If they were university students, they studied
for free, on a course that lasted four, not three years; if they
were farmers they received grants for slurry tanks; if they were
wind farm developers, they found it easier to build turbines.
When they married, almost all the men now wore kilts. The
landholdings were bigger, and the tenant farms much larger,
than in Cumbria. The law of trespass was different, and so were
the licensing laws and the verdicts available to a jury.

The marsh which I had crossed, then, was a frontier between
two identities, if not between two states. Not yet, anyway. If
Scotland votes for independence, we will relearn much older
forms of difference. Perhaps you would not immediately need

to remember your passport, or a new currency. But suddenly, an Englishman in Scotland, or a Scotswoman in England, would be a guest, not someone at home. You would no longer be the responsibility of the other country's embassy abroad. If you were a Scot arrested in London, or English and arrested in Edinburgh, you would be under the custody of a foreign state, a foreign law and a foreign procedure over which you had no say, or vote. Competition between Scots and English in sport would have a different context and tone.

When we faced threats or challenges beyond our shores, we would no longer respond as a single force. We could no longer love the Highlands and London as aspects of a single country. We could no longer criticise each other in the same way, make jokes about our differences, or take pride in each other in the same way.

Which is why I hope the Solway will remain, as it is now, the ambiguous, opaque, tantalising meeting of nations; but never again a frontier to make us foreigners.

19 July 2014

Please come and place a stone on our cairn, at Gretna, tomorrow afternoon. There will be music, poetry, face-painting for children and food for all. If you have a stone, please bring it. If you don't have one, there are stones on site. Everyone's welcome. It's free. All you have to do is turn up.

We are calling the cairn the 'Auld Acquaintance'. We are building it, with stones from around the country, to demonstrate that Scots, English, Welsh and Northern Irish care about each other and are still committed to building things together.

Trying to organise the cairn has been a glimpse of just how much is involved in any 'event' in Britain, from an agricultural show to the Bampton sports day. It's not simply the 'risk management', planning regulations and the rest. My Blackberry packed up by 4 p.m. today, its battery run down by what I think were seventy phone calls in nine hours. But the real work has been done not in emails but by volunteers, working past midnight, on a hundred separate tasks.

Some of it has been artists' work. Paul Jakulis has given days of his time to designing and drawing the cairn (modelled on the Neolithic structures of Scotland and Cumbria). Steven Allen, one of Cumbria's and Britain's best wallers, has been on site repeatedly, working out how to build a dry-stone inner chamber, fifteen feet in diameter, in the very centre of the cairn.

But it's not all architecture. Steve Chettle called this afternoon from a layby to report on twelve tins of white paint, grass-suppressant and fencing. (He has asked me to ask whether anyone has a spare wheelbarrow – we need four.) Lucy Cavendish called from Burlington Stone to give us 200 tonnes of Cumbrian slate; Jim Lowther has offered great boulders from near Penrith; Philip Howard is sending 20 tonnes of collapsed dry-stone wall from Naworth; Norman Harrison has offered limestone.

And it is not just Cumbria. We are getting limestone from Kent, more limestone from Wales, a little red sandstone from Perthshire and a lot of granite from Inverness. (But none of it seems right yet for Steven's chamber walls.)

How do we pay for the transport for any of the stone? Constance Wyndham has set up a 'crowdfunding' website. Yesterday thirty people made donations online between £3 and £100.

At the weekend, we were getting stones and support from local dairy farmers. Yesterday, it was soldiers. A Scot has just written in from Saudi Arabia saying: 'I propose to bring a stone from my golf club (Royal Mid-Surrey) where the remains of James IV are said to be buried in the Priory that was on the site after the Battle of Flodden.'

Today, it was historians and writers Simon Schama, David Starkey, Alain de Botton and Max Hastings. On Sunday, I am hoping it will be mostly families and children from Cumbria and Scotland. Then there is the mystery of websites and social media. Angus Aitken is to be found sitting opposite me, as I write, at a kitchen table, gazing at a screen on which Chris Land is editing a video.

Tara is pondering hashtags. Shoshana is checking the emails I am sending out to friends. Every few hours I fire something boldly into the internet. Mostly the tweets sink without trace.

A journalist has just called, desperate to know how many people we are expecting. I have no idea. Does a cairn appeal? Does the issue appeal? I believe people must – as the referendum gets closer – realise that this is the most important decision we have made in three hundred years. There are a dozen reasons to be concerned. I am particularly struck by how our four nations, peacefully combined and cooperating in a single country, is a fragile miracle. The pressures for disintegration – to break into ever-smaller units, arranged around an ever-smaller definition of your identity set against what used to be your fellow countrymen – can seem irresistible.

That was the story of much of nineteenth- and early twentieth-century Europe. That is part of the story of Iraq and Syria today. But in Britain, we have grown from that creative friction between the different nations, their different literatures, different histories, different politics and characters. It is a country which we, as separate peoples, have nevertheless built together.

But so many are still hanging back from the debate. Some English apparently feel that Scottish independence is too complicated, too political, and nothing to do with them. They are afraid of being seen to 'tell other people what to do'. Sometimes there lurks something worse than insecurity: indifference. Scotland, England, Wales and Ireland have drifted apart over the last thirty years, in a shocking way. We don't study each other's history in schools, we don't really try to understand each other's cultures – it sometimes feels as though we are hardly aware each other exists. Even if Scotland does not separate, we need to become reacquainted with each other.

That is the real reason to come and put a stone on the cairn. Amidst all the economic arguments and political ranting, the cairn is an opportunity for the public, not politicians or celebrities, to show that when our country was under threat, we were prepared to say we cared about each other, that we were determined to continue to build together.

I was appalled by the national campaign for the Union – it seemed tone-deaf, desiccated and determined to reduce emotion, history and identity to economic statistics. Our local campaign to build the cairn appeared on the front page of the Times *newspaper on the eve of the vote. But the former prime minister Gordon Brown's powerful, emotional speech had much more impact. People voted to remain in the United Kingdom by 54 to 46 per cent. Unfortunately, Number 10 seemed to believe this vindicated their dry and distant campaign, and they repeated the same playbook in the Brexit campaign.*

5 November 2016

I have spent a lot of the last four years thinking about my father, and writing about him and about walks through Cumbria, particularly in a book published last week, called *The Marches*.

I thought – at first – that I could learn more about him by interviewing him. Often, therefore, when we sat down for dinner at home, or travelled together abroad, I would put a tape recorder on the table. The tapes preserve his deep baritone voice, with its rolled Scottish 'r's', speaking patiently and at great length about his time as a soldier, and later as a colonial civil servant and intelligence officer.

But I didn't learn what I expected from these interviews. I had known, for example, since I was a child that, before I was born, as a British diplomat in Burma, he had kept a honey bear. Every day when he came home from work my sisters would say, 'Daddy, Daddy, the honey bear is stuck up the tree,' and he would have to climb up to coax it down with a baby-bottle.

But I knew nothing about what he had thought or felt about his life in Rangoon. So I took him all the way back to Burma – after an absence of fifty years – and found his house, unaltered. Entering with him, I expected a sudden burst of new memories. But he was reluctant to leave the garden, where he had found a blackened stump. 'That,' he insisted triumphantly, 'was the tree on which the honey bear sat – "Daddy, Daddy, the honey bear . . ."' And that was all I learned. It was even worse if I tried to ask this talkative man about the character of his brother – who was killed in the war – about his ambitions and frustrations, or his nationalism. All he ever said about his own father was that he was 'a quiet, good-looking man, always reading the newspaper'.

But as I continued on the walk described in the book – from my cottage on the back of Ullswater, over Blencathra, to Maryport, up to Silloth, east to Wigton, fording the Solway to Annan, and then working my way along the border-line to Berwick – I realised my father was not the exception.

During my long walk, almost every one of the hundred or so people I interviewed left me as bewildered as I had been by my father. They spoke fluently about subjects in which I had not expected them to be interested, and were often taciturn about their local area, which I thought would absorb them. I found it very difficult to guess much about anyone.

The man in Jedburgh, playing Border ballads on a Border bagpipe, turned out to sing in an American accent, and came from Essex; remote hamlets that I expected to be filled with farmers contained IT consultants and aromatherapists; and even Willy Tyson, a Herdwick shepherd by Blencathra, who could count fluently in ancient Cumbric, wanted to talk mostly about the time he rode a motorbike to Afghanistan.

It was often difficult to create a coherent picture of an individual's identity. A woman told me that she was a Scottish nationalist because of the miners' strike, and then conceded that most of the miners affected had been in England, not Scotland, and she hadn't known any miners personally. She felt Scotland needed independence because England didn't

understand rural areas; but had herself grown up in Living-
ston New Town, a place with a population larger than the city
of Carlisle, close to Edinburgh; and concluded by saying that
Scotland was a gloomy place, and that she would much rather
live in California.

Which brings me back to my father. His fiercest identifi-
cation was with his Highland regiment, the Black Watch,
with which he fought in the war. He was an extreme Scottish
extrovert, swathed in tartan, serving haggis aggressively to
his English guests, while remaining a fervent believer in the
Union. He invested most of the last twenty years of his life in
planting trees and constructing earthworks around our house
in Scotland.

But if I tried to question him too seriously about any of this,
he would laugh. He never took the trouble to learn the names of
many of the trees that he planted; and although he gave names
to his earthworks (a 'lochan' or a 'ha-ha', a 'duck pond' or a
'dyke'), he often demolished them the following year. When
I questioned him about a new kilt, he said that he had no idea
what tartan it represented, and that he had bought it for £10 in
a charity shop in Crieff.

So I began to see his Scottish unionism, like the woman's
Scottish nationalism, not as a detailed historical claim, or something steeped in organic roots in a particular soil, but instead as
something curiously improvisatory, even whimsical, but which
nevertheless produced a strong sense of national identity. And
I began to see that asking questions about my father's past life
was not the right approach. (Nor were questions about his
beliefs: 'Do you think about death, Daddy?' 'Can't see the point
in that.') What mattered about his identity did not exist in a
philosophy, or in what he had once done: it erupted in present
activity.

What he might or might not have felt on a particular day in 1958 in a house in Burma was irrelevant. He was instead that living ninety-year-old who struggled cheerfully out of bed, straight onto his quad bike, to dig a hole; and he was the man who grinned like a cheeky seven-year-old when I asked precisely what that hole might be for.

HERITAGE LAND

H ERE – HAVING BEEN DISTURBED by the amnesia and the thin, self-defeating identities in the Scottish referendum – I turn to identity and heritage within the constituency. Tourism was the largest part of the local economy, generating billions of pounds of income a year, and much of this tourism was based on the heritage of Hadrian's Wall and the Roman frontier. I knew many farmers, walkers, photographers and archaeologists who were mesmerised by the Roman wall. Their perspective blended erudition with memories of wet boots, the sound of curlews in May, the scent of wild garlic, and trips with their mothers-in-law.

My own tangle of historical facts, speculations, anachronisms and personal memories was summed up in the fort that once stood on the low mound rising out of the wet, reed-fringed sheep pasture at Bewcastle. This was a place near which my friend Steve kept his British Blue cows, and Trevor Telford his sheep: a place which had a problem with mobile signals, whose primary school was under threat, and where there was a nineteenth-century joke about the absence of Armstrongs in the graveyard ('They were all hung as thieves in Carlisle'). It was a bizarre Roman fort – bizarrely located, bizarrely shaped, bizarrely manned and bizarrely resupplied.

Two words on an itinerary copied in a library in Constantinople, and preserved in the Vatican, reveal that the Romans once called this fort 'the shrine of Cocidius', meaning that they had built it on the holy hill of the old god of the Carvetii – that nation whose boundaries almost coincided with the constituency. The Romans seem to have designed this strange fort to keep the shrine of Cocidius and its divine power within the military frontier, as a way of colonising a sanctuary: a tribute to fear of the sacred. Some echo of this fear may have even persisted when an Anglian obelisk of assured craftsmanship and mysterious Christian meaning was driven like a stake into the heart of the site three hundred years after the Roman fort was overrun.

But these were the romantic speculations of a walking Member of Parliament – not archaeological certainties. Perhaps the bizarre anomaly of this fort in the north of the constituency was simply some bureaucratic mishap, or a construction project authorised by a corrupt Imperial freedman.

Meanwhile, professional archaeologists continue to discover and record the smallest engineering decisions of the Romans – the precise turf they selected, their surveying lines, their bridges and their sinuous dance with crag tops. The tourist authorities, the heritage charities and the local councils sponsor roundabouts, reconstructions, quarries, tombstones, trails, visitor arches, pictures, maps and virtual imagery. So that the wall is a joyful, ingenious, scholarly, ironic, ever-evolving memorial: reanimating the mute and resistant masonry for tourists.

Few of my constituents, however, seemed to feel, or at least to express, this form of interest in the history of the landscape around them. Even the small farmers – the most stubborn links to the older traditions – often did not seem to express much sense of connection to the Cumbria of their grandfathers, let alone to the First World War.

16 April 2011

This Tuesday I learned that Wigton is about to celebrate its 750th anniversary as a market town. I am really looking forward to the event, but I have to confess I think of Wigton in 1262 as a place foreign in almost every conceivable way. In 1262 most of the people in Wigton spoke a dialect heavily influenced by northern German.

The lord of Wigton would have worn chain-mail and spent much of his day prancing and practising with his sword and lance while speaking Norman French. In 1262 some people will have regarded Wigton as a Scottish, not an English, town. Others who spoke almost pure Norse would have been more aware of the sea connections to the Isle of Man, the Orkneys, Ireland and Scandinavia. In these and other ways, the Cumbria of 1262 has as much to do with modern Cumbria as the Paleolithic rhinoceros excavated in Bloomsbury has to do with modern London.

The Stone Age monuments of Cumbria seem more alien still. King Arthur's Round Table has become a suburban lawn beside a roundabout at Eamont Bridge; and the deep, secret hollow of Mayburgh Henge echoes with the roar of the M6. The 100,000 man-hours and the miles of sweat and pain that dragged those great boulders across the fells, the endless, repeated rituals and sacrifices enacted on and around Long Meg and her sisters, have been reduced to a place to walk the terrier on a sunny spring afternoon.

Often, our history can feel like little more than a gimmick or marketing slogan. As when, for example, we etch 2,000-year-old sheep-counting numerals (yann, tann, tethera . . .) on the glass door of a conference suite; or when a broadband group cheerfully skates over centuries of cross-border brutality by calling itself Rory's Reivers. Thus we tame the violent, uncategorisable past: domesticating it like a lion in a provincial zoo.

21 March 2015

My father first took me to Hadrian's Wall when I was ten. He worked in China then, and I think I expected the Great Wall of China. We had walked along the top of the Great Wall, from the point at which it rose – tall and neat as in any *National Geographic* film – to the point at which stones began to crumble and subside, and the ruins were lost beneath gravel and sand. We saw Mongolian families in great felt coats, lined with sheepskin: their flocks grazing tufts of dry grass in the grey soil. My father encouraged my attempts to memorise Ezra Pound's version of Li Po's poem about a frontier guard:

> By the North Gate, the wind blows full of sand,
> Lonely from the beginning of time until now!
> Trees fall, the grass goes yellow with autumn.
> I climb the towers and towers
>   to watch out the barbarous land . . .

He tried different Chinese dialects on the dromedary driver, to bargain for Ching dynasty cash (he gave such strings of perforated coins as Christmas presents). We returned, eating the sugar fritters which he had bought from the street-stalls that morning.

When the traveller and historian Camden saw Hadrian's Wall in the seventeenth century, it was still in places almost as tall as the Great Wall, and still edged with its own particular peoples:

> Verily I have seene the tract of it over the high pitches and steepe descents of hilles, wonderfully rising and falling . . . [and to turn to modern English] . . . you may see as it were the ancient Nomads, a martial kind of men who, from the month of April into August, lie out scattering and summering (as they term it) with their cattle, in little cottages here and there, which they call sheils and sheilings. I could not with safety take the full survey of [the Roman Fort at Housesteads] for the rank robbers thereabout.

In Camden's day, the wall stood, in places, ten feet high. Grass mounds cushioned the sharp edges of limestone altars, peat lay in hypocaust foundations which once heated Roman bathhouses. Visitors could scratch moss from the stones to reveal inscriptions which gave the first hints of units drawn from Syria, or the death of a Dacian baby. At Chesters, the bearded head of the god of the Tyne was still half-visible, where he had fallen beside the commanding officer's baths.

Two hundred years after Camden, Walter Scott found Hadrian's Wall already much diminished. General Wade had demolished the central section and used it as rubble for a military road. But miniature plants still nestled in the remaining wall-stones. Scott walked along it, wrote poems about it, and proposed to his wife on the monument. He saw the wall as a symbol of empire, romance and decay; of botany, politics and heraldry; and as the source of a flower for his fiancée:

> Take these flowers which, purple waving,
> On the ruin'd rampart grew,

Where, the sons of freedom braving,
Rome's imperial standards flew.
Warriors from the breach of danger
Pluck no longer laurels there;
They but yield the passing stranger
Wild-flower wreaths for Beauty's hair.

But when I first saw Hadrian's Wall, with my father in 1983, there were no wildflowers left in the crevices. I was not aware at the time that the Department of the Environment had poured a cement core into the wall-stones, and laid a thick concrete cap over the top. The varying colours of the Roman mortar – each recording different periods of Roman restoration – had been replaced with Portland cement.

We walked up a green lawn in the rain. Archaeologists had 'presented the fort' by laying out a pattern of low stone blocks on a sloping lawn. Fallen stones had been lifted from the surrounding area and replaced neatly, so that the fort no longer looked like a collapsing ruin, but instead – in its squares and rectangles – like the lines of a playing-field for a game whose rules were now forgotten. The wall itself was three feet high and lay in the centre of a grass lawn. There were no herdsmen; no sign of any living community or humans, except four tourists in anoraks, and a ticket-seller at the gate.

Signs insisted that it was forbidden to walk on the wall. Further west, the wall simply disappeared. This was not because it was not there. In fact, the Department of the Environment had uncovered an impressive section of red sandstone wall, five courses high. But it had reburied it under a thick layer of grass and turf – 'protecting' it by hiding it. The forts had been stripped bare, their statues and treasures removed to distant museum storage vaults in Newcastle or London, the ground levelled, leaving only the stubby plans of buildings, filled with a thin surface of gravel, and fronted by plastic signs.

This was not always our approach to ruins. I have been reading a book by Christopher Woodward, which explains that in the nineteenth century, Italian families still camped on the terraces of the Colosseum and rode horses round bonfires in the arena. A botanist counted 420 separate plant species then growing in the crevices of the Roman walls, including rare flowers from India or Africa, imported in the droppings of exotic animals brought by the Romans for the gladiatorial fights. Generations of poets and painters flocked to see nature slowly overwhelming the crumbling masonry.

Then, in 1870, the Italian government sent in archaeologists to weed-kill and render every stone, clear out the local families, lock the gates at night, and expose the sewage system. The Colosseum was made as clean and safe as we have made Hadrian's Wall. But who would write a poem about either today?

14 February 2015

On Friday, I came home on the train, feeling that Parliament, and even Britain, was out of sorts. But twenty separate meetings – starting in Penrith, on to Wigton and finishing at a memorial service for Mary Burkett in Carlisle – changed my view.

First, we broke the foundations for the new Sunbeams music centre – a clean-lined, modern building framed by the distant ridges of the East Fellside. (Annie and her team had managed to raise almost £2 million for their music therapy.) Then, at Wigton, I saw the foundations of a factory which will make the £5 and £10 banknotes for the Bank of England, creating more than a hundred new jobs.

One community group had saved the Wigton swimming pool; a second group had a plan to restore the beauty of the Georgian market town, one shop front at a time. Four boys, who had just completed a course in documentary-making, showed me a fifteen-minute film on their experience in foster care, which was spare, subtle and captivating.

Next door, schoolchildren had taken the 120 names from the war memorial and were plotting these soldiers' homes on a map of Wigton, connecting the almost unimaginable scale of the killing to their own local history, and to houses which they knew.

The people behind these projects included civil servants,

businesspeople, charity workers: people who were paid and people who were unpaid. A seventy-six-year-old was leading on the market town scheme; a thirty-year-old woman was the project manager for the £20 million factory; a seventeen-year-old directed the documentary.

Some were exhausted by the relentless modern demands for plans, regulations and funding. And yet they felt pride: in some cases, in buildings which will last beyond their lives; in other cases, through linking people to the past; or, in the case of the documentary, or music, through investigating the present moment. And although many of these people were born in other places, all were proud of being Cumbrian.

Then I went to Mary Burkett's memorial service. Two years ago, she had heard that Shoshana and I were walking from Ullswater to Maryport, and invited us to stay. I had not seen her house before and when we walked up the long avenue of trees, it did not seem real. One end was formed of a stocky, almost squat, pink-plastered medieval pele tower, abutting a long facade of miniature multi-paned windows, apparently unaltered in five hundred years. Mary stood at the threshold with a smile that revealed every tooth. It was difficult to know how tall she had once been, but now she was tiny, dressed in rubber-soled shoes and a striped skirt.

She was living alone at the age of eighty-eight in an unheated house that seemed to consist of a hundred tiny rooms, many unfurnished or decorated only with Central Asian felts, which she had collected on trips around Afghanistan. Having guests must have been exhausting for her, but she did not show it.

At supper (we went out since she kept no food) she described trying to get an English archaeologist, who wore a watch-chain and a three-piece suit, out of an Afghan jail in 1978. She talked about Kushan jewellery and about Lady Anne Clifford, who, as a single, widowed woman, had restored so much of Cumbria in the seventeenth century. And she put us up that night under six felts on a horse-hair mattress.

The next morning, she invited her friend Julian, a painter, to walk with me. I looked back at her, tiny against the facade, on that ridge above the river and the Jacobean garden, and wondered who would ever have the confidence to inhabit such a house after she was gone.

Melvyn Bragg, in his address in Carlisle Cathedral on Friday night, made Mary a symbol of the ancient kingdom of Northumbria–Cumbria. He described how for more than twenty years as the director of Abbot Hall she had built the staff from four to forty, had discovered seventeenth-century and contemporary Cumbrian painters, and championed Cumbrian poets like Norman Nicholson.

Professor Rosemary Cramp, who is almost Mary's age, read from Alfred Wainwright's description of Stickle Pike; and she was followed by a description (by Robert Byron) of an eleventh-century tower in eastern Persia; and a provocative reading from the *Wisdom of Solomon*. And then the Bishop of Carlisle wove her life into concise, and striking, moral lessons.

I sat at the very back of the cathedral. It was night behind the stained-glass windows, and the blue ceiling, set with suns and stars, stretched on into candlelit darkness. Every seat was taken. Some people had come with all their children. I had begun the day feeling the world was fragmented; I ended it linked to the people around me – to Henry Howard, for example, in the choir-stall, whose voice I recognised from when I first heard him sing twenty-nine years ago.

Along the line of tall wooden thrones that formed the canon's gallery were Cumbrians who had travelled not just from the other end of Cumbria (Hugh Cavendish had come from

Cartmel), but from London, and before that Tunisia. Henry Howard's family had sat in the front pews of the cathedral before the Reformation. Others, like Mary herself, were originally from outside Cumbria. Henry Howard's family were ancient Cumbrians who had sat in the front pews of the cathedral before the Reformation. Others, like Mary herself, were originally not Cumbrian at all. They had become Cumbrian through choice. But Mary had half-created through her life, as much as through her work, an idea of a Cumbrian civilisation which bound them all together at the moment of her death, forming the community who had come to remember her on that night.

6 October 2018

Cumbria's excellent Museum of Military Life at Carlisle Castle has a new exhibition, 'Lest We Forget'. I went expecting to focus on the victims of war; I left, however, reflecting on how we fail to remember heroes.

The First World War is less visible in Cumbria than in other parts of Britain. While the centre of every village in Oxfordshire or the Highlands seems to be dominated by a cenotaph or a statue of a soldier, there is no equivalent in Penrith, or Brampton, or Longtown.

The Carlisle museum itself was filled with beautiful Cumbrian memorials to the First World War in marble and bronze, silk and silver. But they were no longer in the towns or villages for which they had been made.

Many of the museum's exhibits came from public buildings or squares that have since been demolished, from tanks reduced to scrap, and even, disturbingly, from finds on rubbish tips. While some of these losses are understandable – it is not easy to maintain an Edwardian drill hall – some of the losses were driven not by necessity but by something that felt more like indifference.

I was particularly taken with a great, engraved silver plate awarded to the best runner in a Carlisle school, made at great expense in memory of one of the war dead – with sixty names around the rim and space for many more – which, for no apparent reason, the school just stopped awarding in the 1980s.

This indifference was particularly striking because almost every case in the main museum re-emphasised how tight the connection had once been between Cumbrian society and the military. There are good photographs of many of the Cumbrian soldiers – taken before their departures for the Western Front – in which the officers look like matinee idols: tall, beautifully dressed, with bristling moustaches; while the men often look like awkward, underfed schoolboys. You could sense the class distinctions embedded as much in the different diets, as in the costs of different haircuts and different cloth for uniforms.

But the photographs also suggested how class distinctions were beginning to erode. The photographs of both officers and men were the same size, in the same frames, by the same photographers. This was the first period in history in which the names of soldiers, as well as officers, were honoured on the same memorials in identical font and in alphabetical, not rank, order.

The hundred people who contributed so enthusiastically to the giant silver shield for shooting must have expected that it would continue to be treasured and awarded for many generations. The officers' mess of the Border Regiment had been persuaded by an entrepreneurial London silversmith (whose advertising brochure survives) to buy a silver statue as a centrepiece for their table at a cost in today's money of perhaps £20,000. They would have hoped that the regiment would admire the piece for centuries to come as part of an unbroken military tradition, which in their minds stretched eight centuries back in time. The piece they chose in the twentieth century represents, in shining armour, lances and rearing horseflesh, the capture of the Earl of Gloucester at a battle near Winchester in 1141. (I cannot see the connection to Cumbria, and the museum did not explain.)

This removal of monuments from communities and associations and public space – and their relegation to exhibitions in museums – has happened very quickly. Here in Cumbria the most dramatic memorials are now the First World War memorial hospitals in Brampton and Wigton, which somehow remained open when their equivalents elsewhere were closed, many only twenty years after they had been opened.

But they too are now squeezed by the modern world: by funding pressures in public health, by new fashions in medicine, by the idea that such heritage buildings are 'no longer fit for purpose'.

Even in the museum, however, these memorials continue to challenge us. It is tempting to dwell only on the waste, the hypocrisy, the suffering and the failures of the trenches: to regard the class traditions as a scandal, and the pride of these soldiers and their families as a delusion. But these memorials also encourage us to do something more difficult. Could we bring ourselves to acknowledge that there might be something admirable in the sense of pride and honour and duty embedded in these photographs, that they might preserve not just a pitiable past but a society that was complex, changing, confident and self-aware? Brave men, in an understated, evolving Edwardian mould?

Take the picture of the Reverend Theodore Bayley Hardy (there is a book about him in the museum), a schoolmaster-turned-vicar of Hutton Roof, who enlisted at the age of fifty-two, far older than the legal age for conscription. Repeatedly, he walked across the wire into no man's land – sometimes wounded, invariably under heavy shell fire – ministering to the dying and leading the stretcher parties back. He did this, night after night, for two years until he was finally killed.

The people of Cumberland who commemorated him in memorials in Hutton Roof and Carlisle expected us to remember him still.

And look again at the picture of him. Think what you like of him in his riding britches and epaulettes, medals and peaked hat but wonder also what he might think of you.

23 November 2013

Why does Penrith not have a First World War memorial in our town centre? Indeed, why is there so little sculpture at all? The only really first-rate sculpture is the Giant's Tomb in St Andrew's Square. It is wonderful: raw hogback stones, soaring crosses, hints of pagan serpents tamed into Christian devils. But it was set up before AD 950, on the frontier of a now vanished kingdom. A thousand years in the rain has left it largely illegible.

Today, one entrance to Penrith is marked by Kentucky Fried Chicken and B&Q, and the other by an avenue of sheds and commercial hangars. The castle, our oldest building, which once dominated the town, is now roofless and ruined, marooned by traffic and overshadowed by the car park and McDonald's. The Beacon is hidden.

Our only really recognisable landmark is the clock which commemorates the death of a young landowner – Philip Musgrave – in 1861. The critic Nikolaus Pevsner has been very generous to Cumbria: he calls the Bewcastle Cross 'the greatest achievement of its date in the whole of Europe'. But he dismisses the Musgrave clock tower as 'utterly insignificant'.

And I fear that if we tried to produce a statue for Penrith

today, it would be even more embarrassing. Take Selkirk. In 1913 it marked the 400th anniversary of the Battle of Flodden with a statue, which is still, a century later, the symbol of the town. But a local committee decided to mark the 500th anniversary only by planting some flowers at the base of the statue, which they called a peace garden. Once discussed, planned, approved and executed, the peace garden cost more in real terms than the original statue. That is one modern way. Another is to try to be bold – and produce immense ugliness: look at the profusion of giant metal mythical figures down Scottish motorways, which are trying to compete with the Angel of the North. Or we play safe and fall back, as councils have all over the country, by installing over-sized stainless-steel balls.

We have sculptors, of course, in Cumbria capable of producing strong naturalistic works. David Williams-Ellis from Lazonby, for example, has made beautiful bronzes, ranging from dancers, now in Shanghai, to a rugby player in full motion, warding a tackle, now in Llanelli, Wales. Could he or someone like him produce a naturalistic sculpture to commemorate the First World War? And what would it represent?

I was tempted at first by the idea of a statue which echoes the imagery of Thomas Hardy's First World War poem 'In Time of "The Breaking of Nations" '. Hardy does not focus on the battlefield: on guns so loud that they destroy ears with a single explosion; on the night turned into day by flares and flames; on fear, blood, loyalty, or young men dying. Instead, he writes:

> Only a man harrowing clods
> In a slow silent walk
> With an old horse that stumbles and nods
> Half asleep as they stalk.

Only thin smoke without flame
From the heaps of couch-grass;
Yet this will go onward the same
Though Dynasties pass.

Yonder a maid and her wight
Come whispering by:
War's annals will cloud into night
Ere their story die.

We are close enough to see that the horse is half asleep, and to hear the whisper. But the man, the old horse, the girl and the boy in the poem have neither features nor names. There are hidden hints of seasons: 'harrowing' happens after the ploughing, and before the ground is sown with a new crop, so that scene may be in late autumn. 'Couch-grass', however, is best torn up by its roots, and burned, in the spring. So we are in the autumn and the spring. But he does not reveal the century or the country.

I have changed my mind, however, about commemorating the First World War with a heroic statue of a ploughman evoking Hardy's rural eternity. An image which for Hardy was difficult would be for us too easy. Hardy was struggling in 1916 to step out of time and place, to produce a difficult image of peace, when politicians and writers-turned-propagandists were roaring out sentimental slogans about patriotism and honour to justify and cover the failures of war.

In 2013, by contrast, our whole culture is too comfortable with the idea that the First World War was senseless, and that the soldiers were only victims. We find it easy to remember the horse and the lovers and civilians at home. We find it more difficult to imagine the combatants and the way they saw themselves.

Our challenge is to rediscover them as soldiers, as actors in movement, not passive puppets. Penrith needs a monument which is intensely local, which represents real men born in Penrith with individual faces, and which has the courage to present them not as victims, but as they might have wanted to be seen.

26 October 2013

The Longtown munitions depot, one of the least likely memorials to the First World War, has been saved from closure today.
We have preserved two hundred jobs in an area where there is
not enough good employment. It is excellent news. But I had
not realised the story of its threatened closure was quite such
an ancient saga.

It was in 1915 that the Imperial War Office first arrived and
enclosed a patch of good farmland almost twelve miles long
between Annan and Longtown. The original farmers may have
been descendants of border-reiving Maxwells and Johnsons;
perhaps they had even been around when the Middleland was
still an independent nation. In any case, their story ended in 1915.
The War Office moved them out. They took their furniture and
their memories elsewhere. I have not traced where they went.

The War Office then shipped builders across the sea from
Ireland. Nobody knew quite what they were building. But locals
were impressed by their drinking. The builders, it is said, would
collect their silvers on a Friday to bribe the train driver to race to
Carlisle where, in the pubs, hundreds of glasses of whisky were
already laid out on the bar. As the drinking got out of hand,
the government nationalised every drinking house within the
region, limiting their opening hours, controlling their prices.

Which is why, in the year of my birth, our pubs around Carlisle were still managed by the government.

The War Office turned this collection of small farmsteads into one link in a vast industrial system that stretched to Africa and Asia. It was as though the Roman army had returned to the Solway frontier, complete with auxiliaries drawn from all over the empire. A manager was plucked from his civilian work in South Africa; another was posted in from British India. Then, 30,000 women, many from the most remote Highland valleys, were imported to work on the site. At Eastriggs, a new garden city was created, with model churches for different denominations to accommodate them (the streets are still called 'Dominion Row' and the 'Rand'). All this in about eighteen months.

The workers stirred great vats, squeezed, rolled, cut and packed, dyeing their skin yellow from the fumes. Their resulting packages were moved by train and then shipped to Flanders, where they were fired, night and day, at the German trenches. By 1918 they were producing 800 tons of cordite a week, more than every other British factory combined.

Hundreds of thousands of German soldiers were killed by the cordite produced on our land. It was a secret installation, and was placed here because this was one of the most remote and sparsely populated parts of Britain, protected from bombers by the Lake District hills.

After the First World War, the government felt it no longer needed the munitions factories. They laid off tens of thousands, and tried to generate industry in the 1920s by selling the custom-made factories and housing. There was interest from an entrepreneur who thought growing sugar beet might work, but it failed on this wet, northern soil. Nobody bought the housing. So the government dismantled the facility. Like the Romans before them, they divided it into regular strips and settled demobilised soldiers on the land.

Then the Second World War came, and once more our remote, sparsely populated position on the borders made us very appealing to military planners. The army returned, making the depots part of a military infrastructure which sprawled across the north, from the army training area at Otterburn to the airfields at Carlisle and Silloth. When my father first saw Longtown as a Black Watch soldier in 1943, 5,000 infantry were camped along the south road, sleeping in tents or under trucks.

After that war, Iraqi pilots came to train near Walton. The Blue Streak missile site was developed beyond Brampton. The RAF Carlisle site became the Kingmoor industrial park. Eastriggs was abandoned, but because it was contaminated by slightly sinister sounding 'specialist ammunition', it was left empty, and is still protected by guards.

In the end, there was just Longtown. The last few hundred acres out of the twelve-mile-long site, the last two hundred workers out of 30,000. Six years ago, the Ministry of Defence announced that it would be too expensive to upgrade what remained, and that it too, therefore, needed to be closed. So much civil service time had been invested in strategic plans and redundancy schemes, that it seemed almost inconceivable that it could be saved.

But they underestimated Neil Scott, the trade unions' representative at Longtown. Each time the MoD produced another strategic plan, he went through every single document, questioning every figure. He pointed out that they had underestimated the transport costs and risks of moving munitions to Warwickshire. He emphasised the challenges that would follow if Scotland went independent. He warned of the dangers of concentrating all the UK's munitions at a single site. He asked us to consider the possibility that the world might become less peaceful. He chased me, almost weekly, for updates, to see what arguments I was putting to ministers. He drove down from Longtown to my surgeries in different towns across Cumbria, he came to Parliament.

He showed me twice in detail around every ammunition shed. He persuaded me to invite the defence minister to visit the site last year. He encouraged the three neighbouring MPs until we took the argument all the way to the prime minister.

It was a tough fight – the army is shrinking, overseas operations are reducing, and the whole policy was to fall back on Warwickshire. It worked. We have somehow managed to keep the site open, save two hundred jobs, and get another million pounds of investment out of the MoD. And if Neil is right about the direction in which the world is going, Britain may be grateful.

The munitions depot at Longtown is still open and has found a new lease of life in the context of the Russia–Ukraine war and Britain's push to rearm. In October 2024, more than a century after it was first opened, the MoD announced that £86m more would be invested in the site, supporting 450 jobs.

13 November 2010

> That at the hour when the Armistice came into force, the 11th
> hour of the 11th day of the 11th month, there may be for the
> brief space of two minutes a complete suspension of all our
> normal activities . . . so that in perfect stillness, the thoughts
> of everyone may be concentrated on reverent remembrance
> of the glorious dead.

Those are the words of George V, announcing Remembrance
Day in 1919. This ceremony, now seemingly so natural, did
not exist before that moment. Assyrians and Sumerians, for
example, commemorated war through stone carvings of exuberant victory parades. Rome was defined by the extravagant
'triumphs' of its conquering generals.

It took the First World War, it seemed, to change our
monuments from images of conquest to images of sacrifice,
and to transform our memorial from chariots and chants to still-
ness and silence.

My father was one of two brothers. Their father worked
in British India. They were sent to the same Scottish boarding
school aged seven. They spent all their holidays together, often
walking alone through the mountains. And because their
father had served with the Black Watch in Iraq in the First

World War, they, too, joined the Black Watch in 1940. Uncle George was wounded at Alamein and killed in Sicily, where he is buried.

One of my earliest memories is of seeing my father place his thumbs down the seams of his trousers at the sound of the Last Post, and come rigidly to attention, in front of the war memorial at my boarding school. Two minutes of silence seemed a very long time for an eight-year-old.

My father first saw me stand to attention at Remembrance Sunday when I was eighteen. I was in my Black Watch uniform, with my poppy in my lapel and my sword drawn, and I was very nervous. I had practised punching my sword out to the right, for the salute, so often that my arm seemed to remember the action by itself, but when the bagpipes sounded, and the march began, my steel-soled shoes slipped and I almost fell on my back, kilt in the air, at the head of a company of soldiers.

It was 1991 and the Black Watch had last gone to war in Korea forty years earlier. During the two minutes' silence, I looked through the rain at the veterans, like my father, who understood war as we did not. The Cold War was over, and our training felt as though we were rehearsing for a play which would never be performed. But we were wrong.

Twelve years later, my very brief stint in the army far behind me, I was posted to Iraq, and was asked to lay a wreath on Remembrance Sunday in the cemetery in Al Amara. During the two-minute silence I stood in front of ranks of British soldiers in their desert fatigues, looking at a wall inscribed with the names of all the British soldiers who had been buried in that same Iraqi field in the First World War. There were 4,621 buried under our feet, including half a battalion of the Black Watch, one of whom, 'Private Frederick Bewley, S/6436, 2nd Battalion', was listed as 'born in Langwathby, Cumberland. Son of Elias and Mary Bewley, of Ivy Cottage, Langwathby'.

Every standard detail of every Remembrance Sunday was played out under the Iraqi sun: the poppies, the march with medals, the Queen, the words 'at the going down of the sun . . .', and the C and G notes repeated three times by a lone bugler to introduce the Last Post. But, at the heart of it, was silence.

At the end of the First World War, where an ancient Greek would have built a hero's statue, we built a tomb of an unknown soldier; where an ancient Greek, or even a modern American, might have a funeral oration, we have silence. We sense, perhaps, something in death which goes beyond words or images or individuals, which our minds can only equally, anonymously and singularly confront in silence.

My friend Nick, who joined the army four years after me, has served in the Balkans, Baghdad and Kabul. Today he is in Helmand. Tomorrow he will be standing in silence with his brigade in a province where more than three hundred British soldiers have been killed. There will be silence in Canada, whose troops have been in Kandahar, and in Australia, whose troops have been in Uruzgan.

And it will be a silence shared by civilians in Vancouver and Sydney and Penrith. And it is a silence that we can recognise in the very first time it was introduced, ninety-one years ago (described in the *Manchester Guardian* of 1919):

> The tram cars glided into stillness . . . the mighty-limbed dray horses hunched back upon their loads and stopped also, seeming to do it of their own volition. Someone took off his hat, and with a nervous hesitancy the rest of the men

bowed their heads also. Here and there an old soldier could be detected slipping unconsciously into the posture of 'attention' . . . Everyone stood very still . . . The hush deepened. It had spread over the whole city and become so pronounced as to impress one with a sense of audibility. It was a silence which was almost pain . . . And the spirit of memory brooded over it all.

Crieff, 5 September 2015

My father died two weeks ago, at home in Scotland. He often mentioned that he might die.

An hour before he died, I was reading to him, and he said gently he would prefer to talk about what I was up to. I talked about what made me proud of Britain, and in particular Cumbria, today, and he talked about his pride in 'the Empire', by which he meant his and his colleagues' work in Malaya, Singapore and Hong Kong after the war. ('We left them better than the mother country.')

Then he asked to rest. Ten minutes later he sat up. He gave me clear, concise instructions: 'Take off my jumper. Lie me down. Right, roll back my tongue.' There was no sign of any fear of death. He remained practical, calm and focused on the matter in hand, until the moment he lost consciousness. I could not get his heart started again.

I was not sure what kind of funeral to give him. He liked jokes. His favourite author was the Cumbrian George MacDonald Fraser, and his favourite MacDonald Fraser scene was one in which a Highland regiment in North Africa distracted a pompous general, who was about to fail the regiment in its inspection, by gathering a group of passing Bedouin to join, not an eightsome, but a hundred-and-twenty-eightsome reel.

Sometimes my father could seem more like that grandiose general. He had been proud that in his ninety-third year he had published two books, had his coat of arms hung in St Paul's, had a military portrait painted of him for the Royal Collection, and had been awarded a Légion d'honneur, as a Normandy veteran.

But the funeral instructions he left weren't pompous. They were open and relaxed, emphasising that the family could do exactly as they wished, though if they were short of ideas, he had some playful ones. He wrote:

Apart from the death of my beloved elder brother, killed in action, there have been no serious tragedies in my life. Wonderful family. As to my faith; it is a simple one. The Church, and its music, have been part of my life for as long as I can remember. Hereunder is my formal contribution to the Stewart file. 1) I have no strong views on the subject, but all things being equal, and legally possible, I should prefer to be buried with my dogs, at home. 2) If, however, Sally and Rory and my other children would prefer something different, so be it. 3) I am disinterested in whether I am cremated or not. Once again the views of my family, not mine, should prevail.

Memorial or funeral service: The form is not of importance to me, but I hope that tears will be minimised and jollity will prevail. If there is to be a formal farewell, I should like a piper to play a traditional lament, followed by a bugler to play the Last Post, particularly in memory of my brother and all my friends who were killed in action in the war. The wake: plenty to drink, alcoholic and non-alcoholic, and plenty to eat (finger food, sausages, high class sandwiches, better still, the Robbs and their spit). Perhaps we can borrow my old school for service and chapel.

To sum up, gaiety, even perhaps song and dance, should be the hallmark of the exercise. If the weather is fine, a farewell Strip the Willow, Eight-some, Dashing White Sergeant and Gay Gordons would not tax the young or the old.

We began with a service in his school chapel. His brother's photograph had been laid in the glass memorial cabinet by the door. It showed him exactly seventy years younger than my father, in his Black Watch uniform. The photo was taken before he was ordered to Sicily, where he was killed. Then we drove through the Highlands – the sun was on all the slopes of the Sma' Glen – past the Roman fort, which my father and his brother had helped to excavate as schoolboys in the 1930s. We ended at home.

Six of us lifted the heavy coffin and, led by the piper and three clergymen, we carried it down an avenue. We were followed by my mother holding his grandson, Sasha, and my sisters, Heather, Annie and Fiona, and a hundred mourners, some in kilts and blue bonnets, some in hairy tweed. The ladies' high heels sank into the mud. There, on consecrated ground at the edge of the garden, we lowered him into a deep grave.

Two buglers played the Last Post. Then the Black Watch piper played the regimental lament. I held my nine-month-old son in my arms. His legs danced to the pipe music. And the piper led us back with my father's favourite regimental marching songs, 'Black Bear' and 'Hielan' Laddie'.

We had, as my father suggested, the Robb's roast pig on a spit. The sun shone on the nearly thousand trees that he had planted in the field in front of the house. His dog came out to watch. And then the piper played the eightsome so that we could dance. My father would not have approved of all of the steps, and it was certainly less orderly than he would have liked. He probably wouldn't have recognised it as an eightsome.

But there on the dry grass were Cumbrians and Scots, colleagues from British Intelligence, his family, Black Watch
generals, my friends, his friends, young and old, spinning and
twirling and laughing, arm to arm in the most traditional of the
Scottish dances, on his front lawn, performing in his honour, not
an eightsome, but a fifty-seven-some reel.

NO MAN'S LAND

M ANY OF THESE PREVIOUS LETTERS are about Cumbria as a frontier, a place which, being on the very edge of England, was as far as it could be from central government.

But the constituency also featured in history not as a border, but almost as a blank space on a map, a place where things could be done which could not be done elsewhere. And when governments discovered that the place was not in fact empty, they often tried to empty it.

The Romans, as always, were the first. Our local Carvetii nation was poor, lived in scattered farmsteads, and trod lightly on the earth – even today, heroic efforts by archaeologists have recovered only fragmentary signs of their existence. Julius Caesar said that the tribes of the north-west, such as the Carvetii, did not even plant crops. In truth, we can see that they did plough the fields and plant crops. But the Romans built a wall that cut straight through their plough lines, severing their fields. And the legions perpetrated what was almost an ethnic cleansing, demolishing homes and farms in a large area north of the wall. In order to forge the infrastructure for an imperial frontier – and the fears and fantasies which justified it – they made it a no man's land.

The independent Middleland nation which re-emerged in this space after the Romans left is recorded largely in myth – this is the centre of many of the stories about Merlin and King

Arthur – and in the Cumbrian epic, which describes the killing of drunken warriors at Catterick. It became a space into which expanding Germanic-speaking Angles drove stone crosses to mark territory and boundaries. The Anglians eliminated almost all the place-names on the east side of the constituency, replacing them with their own (like my hamlet of Butterwick, near Bampton). And Norse settlers such as the eponymous 'Ulf' seem to have eliminated all the names which had been used by the people who preceded them in the west – hence Ullswater, which means 'Ulf's lake'. Then what remained of the old Cumbrian kingdom was squeezed to death between the English and the Scots.

And the English and Scottish kings turned most of the constituency, and the border regions of Scotland, into royal forests: it became a zone for trees and animals, not humans, in which it was forbidden to farm, hunt, deforest or build. Another no man's land. Did the Scottish and English kings impose these forests because they already saw it as a blank space? Or was the assertion of wilderness an excuse to give themselves emergency powers on the border not granted by the constitution – using forest law much as the Romans used military law, and their successors used Border law, for strategic control of our area? And did similar reasons underlie the decision to put much of the remaining land in the twelfth and the thirteenth centuries under control of the monasteries? The Church, too, presented this area as a wilderness – and therefore particularly suitable for monks who were seeking places away from men.

From 1300 to 1600, our land became the Marches – governed under its own Border law almost as a foreign colony: the place outside the law celebrated by Walter Scott. In the early nineteenth century, it was again celebrated as a natural wilderness, half-perceived and half-created by Wordsworth. In the twentieth century, it was considered a suitable location for munitions factories, for the nuclear plants, for nuclear waste, and for the Blue Streak missile facility. Then it was proposed that whole valleys should become reservoirs, or be drowned under dense plantations of commercial forestry. Again, outsiders seemed to treat Cumbria as a blank space where things could be done which would be resisted in other, wealthier, more central, more connected parts of England or Scotland.

Then, in 2010, the rewilding movement arrived in Cumbria with the force of environmental necessity behind it: justified by climate change, biodiversity collapse and carbon sequestration. Again, the aim was to turn much of the constituency into a wilderness, free of human farming and human management. Sheep would be removed from the fells. Beavers, lynx and wolves, extinct in Britain for centuries, would be imported from other countries and released. Nature would be left to regenerate itself. The science was sometimes compelling, the funding substantial, and the vision appealing to urban sensibilities. But the implementation would often mean the end of farming communities that had shaped the landscape for thousands of years. A tension emerged between two radically different ideas of what the countryside should be: a working landscape supporting human communities, or a managed wilderness, a blank space, optimised for carbon storage and species diversity.

30 April 2011

Last week, I was able to walk from my back garden at Helton to Windermere and Wastwater, and loop back to Keswick. It will be a long time before I forget the great processional ridge of High Street, climbing from cairn to cairn, the deep lanes below Troutbeck buttressed with slate as tall as tombstones, the white blossom on the bare black hedgerows, the pink tendrils on the sycamore. I was walking with Shoshana, and I was proud to be able to show her Cumbria without rain.

But the fierce sun turned the scree around Wasdale Head into a bare, dusty blaze of gravel, and on the great stone steps climbing beside Rossett Gill (as well-laid as a Hindu pilgrimage trail) the heat was like the heat before an Indian monsoon: I could imagine a temple by Angle Tarn, with floating ash and a contemplating saddhu.

I loved standing on the shoulder of Pillar, gazing at two frayed white clouds in a blue sky, but it was not the high peaks at midday that I remember most. Nor the signs of spring: the clown-like muzzles of the Swaledale lambs and the dark shapes of the Herdwicks under their grey mothers, or brown tips of the oak breaking open with leaves as tiny as a miniature Japanese maple. Instead, I remember a low, barren slope with an air of autumn. We were climbing up from Buttermere, on the old pack-road to Keswick. The sun had set an hour earlier. The long, gentle valley was silent. There was no house, nor road, nor even a single track. A sudden late burst of rain had reawakened the shifting light on the hills.

The climb down from the watershed, towards nightfall, was a welcome descent into a fertile plain: with every step, the ground was richer, the trees broader, the cottage gardens more ablaze. Near Derwentwater, April, May and June had all come at once: the apple blossom blazing over two-feet-tall poppy stalks, daffodils beside bluebells and hawthorn, and tulips beside late-flowering rhododendrons. But what I will treasure most is the bare ridge-line, just after sunset, where the grass was sere, and the gill so cold that it felt like one of the last days of September.

7 November 2015

I am writing this on a train on my way back home to Cumbria, after a day as the environment minister listening to ideas for 'a twenty-five-year plan for the British environment'. Everyone insisted that it was not necessary to choose between different competing priorities. Instead, they said, in the meeting, that everything could be balanced.

The more people I meet, however, the less plausible this seems. Some Cumbrians want to achieve food security by producing more food; others want less-intensive farming to support butterflies or increase animal welfare. Some want new wetlands for wading birds; others mountain-bike trails. There are those who want beautiful stone buildings in the landscape. And there are those who want cheaper buildings and believe that this requires fewer regulations. Some would prefer to use the land to generate energy by paying for wind turbines and solar panels, others by allowing coal mining or fracking for gas.

Even when we agree that we want to plant more trees, we disagree on what type. Do we want Sitka spruce? So that Britain can support a wood-processing industry employing thousands? And so that we no longer need to import the wood we use for burning, or building houses? Or do we want to plant native sessile oak – because caterpillars eat the oak buds, bats roost in the oak holes, hundreds of insect species feed on the trunks, badgers munch from their acorns, and beetles and mushrooms thrive on the leaf mould beneath? Or perhaps we should not plant trees at all, and instead conserve the peat to support the great sundew and the heath butterfly, to clean water and prevent flooding, and to trap carbon. (It is currently calculated that the UK's carbon store in peatland equals the entire globe's annual carbon emissions.)

The combination of such views – or rather disagreements – is filtered through people like me in Parliament and government into laws and agricultural support systems, into the billions of pounds spent by water companies, into planning regulations and Brussels-shaped environmental protections. But somehow once these systems are in place, MPs like me, whose predecessors created these structures, find them almost impossible to change. And this inherited framework in turn defines almost every detail of our water, land and even air.

Sometimes, you can see our inconsistent priorities lying starkly alongside each other in a single valley. Take the valley north of Brampton, which I visited last month. Four hundred years ago, it was all deep, treeless peatland, with a scattering of small family farms and tough, squat cattle. Now, very little meat or food of any kind is produced from this land. Many of the farms have become second homes, used only for holidays. There are, however, straight-edged conifer woods where the Forestry Commission began to plant dense commercial softwood in the 1920s. There are concrete roads and the landing strips where the Ministry of Defence placed a ballistic missile test site in the 1950s. In the south, Natural England preserved a strip of ancient woodland on the River Lyne in 2000, which the Forestry Commission in 2012 supported as a place for wood cabin tourism.

In the centre is what remains of the original peat bog, which was destroyed in accordance with one set of government incentives in the 1960s, at a cost of millions, to comply with a new European directive. And that is before you turn to the new housing, or the proposals to build wind turbines, mobile masts and fibre broadband. The farmers we admire, the trees we adore, the landscape we love, somehow have to survive on a small island, amidst the tumultuous juddering intersection of different general government priorities changing every decade, and the pressure of 60 million individual votes.

Over the last century, many things have improved. We have embraced brilliant innovations which have totally transformed our productivity, allowing us to feed fifty times the number of people from the same acreage. We now have the satellites, the DNA testing, the research tools and the models of natural capital which can allow us to understand the environment and the impact of our policies in a way that was unimaginable even a decade ago.

But we still remain cruelly exposed to floods, to droughts, to animal diseases or tree diseases like ash die-back. We have to struggle with potential extinction of species and the final exhaustion of our soils. We face a daily battle against the ugly or the horrifying. It is often difficult to understand how we have an environment at all – how we didn't, like the Easter Islanders, simply cut down all our trees, consume all our natural resources and starve. (Writing this makes me wonder whether the Easter Islanders did in fact do that.)

Now, when I take my Sunday walk from my cottage to Trout-beck, I can see imprinted on the land around me the impact of the twelve different grant structures, and seven different government agencies. The regulations and the subsidies from these bodies have changed the very length and colour of the grass by limiting the number of sheep who can be kept on the fells. They have shaped the types of heather and flower which are now emerging on common land. I can see the mark of their subsidies in the new fencing, the tree planting, the new rushes and the bogs, the repairs to the metal quarries, the lack of dredging in the rivers and the new extent of the floodplains. And most of all I am conscious that somewhere in the middle of it all stands an infuriated farmer, trying to keep doing what his family had done for generations: producing food.

For the first fifteen miles, I do not pass a tree on the fellside. Approaching the ridge called High Street, however, I see a piece of blackened wood embedded in a peat bog: the branch of an oak tree which 5,000 years ago grew from an acorn on this upland. And I look down towards Grisedale, where I know there is a pollarded ash, almost twenty feet in diameter, which may have been there since the Viking invasions – a reminder of the moment when the trees began to be cut to be fed to stock. Then in Troutbeck, at the end of my walk, I come down past young hawthorn and ash and oak, planted up ghylls and gullies, in accordance with 2008 subsidies, beside flocks of Herdwick sheep preserved by Beatrix Potter and entrusted to the National Trust.

And somehow, amidst the cacophony of nineteen different priorities and regulations, we have managed to preserve something which still seems so tranquil, so healthy, so beautiful, so moving. And so natural.

13 October 2012

If a century ago you had climbed the pale, soaking fell-grass out of Bewcastle and passed the Bloody Bush, you would have entered Tynedale: the close-cropped owner-occupied farms of old border clans such as the Milburns and Dodds, whose barns were the converted battle houses of bandits. But now the very existence of the valley can only be guessed.

A rough carpet of black-green needles, trembling forty feet above the ground, conceals the land. Alaskan spruce trees, climbing up to Peel Fell, engulf the stone markers, over-shadow Carter Bar and Reedsdale, and colonise hundreds of thousands of acres of what was once moor, mire and upland farm. And at its centre, with houses drowned beneath the waves, lies a vast artificial reservoir, built too late for industry that no longer exists.

This is Kershope and Kielder. There are delicately laid mountain-bike trails, and in clearings you can hear the great glass-cabined harvester machines, ripping and slicing trees. But I have walked for a whole day, clambering around fallen trunks in the fire breaks, stumbling over the bare ridges and furrows, without seeing a human and hardly a bird.

Almost all of us, instinctively, want to protect woodland. But when we think of forests we tend to think of native trees – the ash of Cumbria, the coppiced hazel and thick-girthed sessile oak under the cliffs and caves of the Eden between Armathwaite and Wetheral. We are interested in the ancient woodland of the Fairy Tables along the Black Lyme, because its soil, undisturbed for hundreds of years, preserves unusual lichen and insects.

But the creation of this woodland involved draining thousands of acres of rare upland mosses, ploughing and close-planting ancient grasslands, and replacing it all with miles of American Sitka. Yet this blanket of almost lifeless monoculture is the very public forest estate that people have been campaigning so hard to save. The independent panel on forestry has just agreed with the campaign; and so do I. Why?

The answer is that in forestry – as in farming – commercial production is vital to sustaining things far beyond Sitka monoculture. Timber-processing brings £600 million a year to Penrith and the Border, and provides employment (through BSW, Jenkinson's and others) for 2,500 people, more than a third of the total number of forestry and wood-processing jobs in England. And it is a good industry. We are learning again that wood is one of the most impressive, renewable products for building houses: it can be felled, replaced and recycled in a way that is inconceivable for aluminium. Wigton's Innovia is showing how wood cellophane can be spun into the most extraordinary films for packing and labelling: entirely renewable and biodegradable, and in a manner that is impossible for plastics.

Employment in the commercial forestry sector continues to draw thousands to develop their skills as foresters, and to study at places such as our National School of Forestry at Newton Rigg.

And this education and skills base underlies the thousands

more who go on to work as professionals or volunteers in ancient forests. The environmental programmes, from planting native broadleaves around the spruce to regenerating the moorland, are often run by commercial forest estates. And at Kielder you can see the Forestry Commission putting more and more energy into bringing visitors into the estate: developing the reservoir as a beauty spot, creating a beautiful bridge for a marathon track, or opening an astronomical observatory for the public.

This is a lesson for farming, too. There – as in forestry – commercial production matters. In both cases, it provides the incomes and maintains the enterprises which people and sustain the landscape. In both cases, government has a role. Trees take forty years or more to grow, and grassland evolves slowly.

These are, therefore, not simple commercial propositions – small changes in commodity prices can produce very damaging long-term results for our industry, our products and our land-scape. I was pleased when the government dropped its proposal to sell off the forest estate.

But let me end with a note of caution. The independent panel report wants to increase by 50 per cent the amount of woodland in Britain. Others want to double the tree-cover in the Lake District, expanding it from 10 to 20 per cent of the land mass. But we have not been a country of limitless Hansel and Gretel forests at any time in recorded history. (We now know that northern Cumbria had been felled, pastured and ploughed long before Hadrian's Wall was laid.) And, despite all the benefits of forestry, we should not allow our land to be turned into a giant forest now.

Our forests are a unique and wonderful heritage, but they should exist alongside, not instead of, other landscapes we love:

open moorland, meadows and upland sheep farms. The interest of the British countryside lies in its balance between valleys, cliffs, woods, fells and pasture. Kielder is a prodigy, Ennerdale is a delight, but we should not set a target which would drown Ullswater or Wastwater under trees.

18 January 2014

Kielder Forest again: the north-eastern edge of this constituency. Banksy's wife's family had originally been sheep farmers
on this spot. The government began to plant the slopes with
trees after the First World War to provide pit-props for trenches
in a future war. Once their pasture, walls and sheds had been
swallowed by trees, the family had become foresters. Banksy's
father and father-in-law had been foresters. So had Olly's father,
brother and son. Banksy and Olly lived in forestry cottages.
Banksy's wife worked planting trees. Their nearest town was
Bellingham. From the bottom of the hill, it was eighteen
miles away.

Now from the Bloody Bush stone, 160,000 acres of dark trees
flow into every hollow in Tynedale. It appears almost as a blank
space on the map, without roads or villages. As a walker, it is
almost impenetrable: a slow push for five hours on steep broken
ground, through the tightly planted trees, parting soft branches
and sharp needles, in semi-darkness.

When Olly's brother began, the men used axes. By the
1960s, they were using the Bowman saw. At seventeen,
Banksy was given a jacket and some steel-toed safety boots,
and at eighteen, a chainsaw, but there had still been hundreds
of forestry workers. At twenty he had rebelled, and tried to

become a farmer. It had not worked out, and he had come back to chain-sawing.

He had worked in a pair. 'I was put next to one of the best chainsaw operators in the North Tyne valley,' he says. They worked in sight of each other, and met every twenty-five minutes to refuel the chainsaws.

'Malcolm told me, "Learn how to sharpen your saw, present your wood and fell your trees: always keep it tidy and you'll make money." I carried a spare T-shirt for the rain. And at the end of the day, I brushed away the branches, to prepare for the next day. I go to a masseur now,' he finished, grinning, 'every two months, for my back.'

Harvesting machines had been introduced when Banksy was in his twenties. Now, instead of hundreds of men, they needed ten. Banksy had moved into supervision. Olly had trained as a harvester operator. 'He won't mind me saying this, he can't understand the computer side of it, but Jason, his son, is up to speed on it, so they work together.'

I watched Olly. He was in the cab of the quarter-of-a-million-pound machine, peering out of the bullet-proof glass. He pushed it forward, over a carpet of branches he had laid, to keep it above the boggy ground. Its robotic arm swung at a ninety-foot tree, grabbed it, twisted it, yanked it forward, stripped its top branches and then immediately began to cut it into sections.

Jason measured the diameter and length with an electronic instrument, and fed the data into a computer, straight to the sawmill. One man could now cut and finish a hundred trees in a day. 'Olly and Jason came at 4.40 this morning,' said Banksy. 'They do an eleven-hour day. Operators are working two or three trees ahead, working out what they can get out of the branches. In the winter, they come in the dark and leave in the dark. They're not very talkative of an evening, because they are brain-dead, just staring at trees all day. Olly's wife works late, so when he gets home, he prepares the dinner.'

'What happens if the tree falls on top of him?' I asked.

'He gets jip off me for damaging the cab.'

Now there was talk of new technology, needing fewer forwarding machines and operators. 'Sons and daughters aren't interested in staying in the forestry industry,' Banksy said. He had decided that his own daughter needed to get out of the North Tyne valley. She lived in York, had a psychology degree and a teaching degree, and was looking for work. The school which Banksy had gone to had closed. They tried to always use the village shop to keep it running, because if they were snowed in, it was the only shop they could reach.

Banksy walked me down the hill to introduce me to his collie. As we came onto the road, an old shepherd pulled up next to us, looked at the dog, and said: 'Ay, you've got one of those stylish dogs!' 'How do you mean?' Banksy asked. 'We got dogs that bark a lot,' the shepherd replied. 'He thought it was a party dog,' said Banksy, when the shepherd had gone, 'but it's not.' Banksy used his dog on his smallholding. They owned a hundred sheep and six cows.

Banksy seemed now to be spending a lot of time commissioning wooden ramps through the forest for mountain-biking. He marshalled fell-running races. He was working to develop patches of wood for owls and ospreys. He talked of the 'dark sky' that they preserved over the forest, and the Forestry Commission star observatory. I wondered whether he disliked this change from chain-sawing to recreation services. He seemed excited by it.

At his cottage, he had changed into his Ron Hill tracksuit. He was going fell-running. He pulled a sled full of sand behind him, because he said the hills weren't high enough here. I said goodbye to him at the reservoir. It had been built to power manufacturing industry in the north-east. The manufacturing had not come. There were now hotels and lake bungalows along its shores. Banksy's father's cottage was somewhere deep below us, buried under the water of the reservoir.

16 June 2018

Rewilding seems to be fast becoming as much a fashion among landowners as Capability Brown parks were in the eighteenth century. It is spreading from Yellowstone National Park to the Cumberland plain.

In Scotland, Anders Povlsen, the richest man in Denmark, is turning 220,000 acres of land – about a third the size of the entire Penrith and the Border constituency – into a rewilding project. In Sussex, four hundred separate dairy fields have been abandoned to nature.

This might simply be a diverting chapter in the history of the British landscape. But increasing numbers of people are now calling for the entire uplands of England – and in particular the Lake District – to be rewilded, and this idea is attracting increasing interest from politicians, government agencies, lobby groups and landowners.

In the Lake District in particular, the idea is often introduced as though it were simply a sensible and moderate way of addressing some of the problems caused by overstocking of sheep in the 1980s – most especially overgrazing, water pollution and methane. And rewilding is disguised in vague statements about addressing climate change, improving biodiversity, creating 'natural flood prevention' and restoring natural woodland. All this is shrouded in disingenuous statements about 'respect for farmers'. But few of us are aware quite how radical the consequences could be.

The ultimate objective of rewilding is to remove all impacts that humans have had on the landscape. Unlike a conventional environmental scheme which relies on carefully scientifically tested interventions (tree planting, river management, seasonal grazing and the rest), rewilding aims to allow animals and plants to restore the landscape by themselves, through natural processes. This in turn requires reintroducing animals which humans removed from the British landscape centuries or even millennia ago: predators such as lynx and wolves, or high-impact species such as beavers and bison, which can control the land-scape 'top-down'.

Rewilding is rarely the most cost-effective, reliable way of restoring biodiversity. In fact, it comes with environmental risks. We are already losing meadows, oak trees and hedgerows, as a section of the Cumberland plain is turned back to bogland, and with them the many species that thrive on pastured meadows and on mature oaks. Until apex predators are allowed to move freely, beaver and deer will prevent new mature trees emerging in any numbers. And if the apex predators emerge, and open areas are replaced by denser woodland, we lose the species, such as hedgehogs, who thrive on the woodland edge.

But some rewilders – I think of Ben Goldsmith, who is on the board of the Department for Environment, Food and Rural Affairs – embrace this risk because of a fundamental emotional commitment to the 'pre-human', carnivore-dominated wilder-ness of the Pleistocene – out of a sense of guilt at the impact humans have had since, and out of a desire to escape what is felt as a tame, safe, modern environment. And they want to work at a very large scale, not by simply creating a nature reserve in one Cumbrian valley, but by creating a vast core of interlocking corridors potentially covering the entire uplands of England.

Many of the apparent environmental benefits of rewilding come at the price of imposing environmental costs on someone else. Clearing sheep from a Lake District valley doesn't mean Cumbrians stop eating lamb; they just begin to eat lamb from abroad. The methane from the sheep continues to enter the atmosphere, and while Cumbrian fellsides may become wet and boggy – more dominated by reeds and mosses – we push our problems to New Zealand or Patagonia, making them desertify under the pressure of sheep, so that lamb can be shipped thousands of miles back to be eaten on Cumbrian tables. Or we crowd our sheep into indoor sheds and give them oil-based products and animal feed grown abroad in what was once rainforest. Why can't we let them eat our rain-fed grass?

But the most fundamental problem is that rewilding was originally intended for places like the Yellowstone National Park – a vast, unfarmed, uninhabited wilderness considerably larger than Cornwall. When applied to our smaller and much more densely populated island, which has been farmed for thousands of years, the consequences are quite different.

It is not simply turning the clock back a few decades to the time before modern farming techniques, but turning it back millennia. There have been farms on the fellside, with pasture grazed by livestock, for at least 6,000 years. The Eden valley had been cleared of its primeval forest well before the Romans built their signal stations.

The most distinctive features of the Lake District – the pollarded ash trees, the dry-stone walls, and the contrasting colours between the more tightly cropped green pasture below and the greys of the wilder common land above – date back to the time of the Vikings, from whom many of our farmers descend. Close-cropped sheep lawns surrounded Shap at the time of the monks. This farmed landscape is very close to the landscape that inspired Turner and Wordsworth.

And unlike the Highlands of Scotland, where the farms were removed in the Clearances, here the small family farms have survived, providing one of the few fragile surviving connections to our historic landscape: a foundation for our tourist industry, the bedrock of our communities and our village schools.

Rewilding is not a gentle return to a natural past. This is not simply because rewilding is sometimes naive about food production, careless of the impact on other countries, blind to the texture and history of our rural areas and its links to our literature and identity. It is fundamentally because rewilding leaves so little place for human culture in the landscape.

1 February 2014

Steve was repairing a dry-stone wall. It was almost eight feet high. I suggested it was unnecessarily high. He agreed that a sheep did not need a wall so high, and did not offer a theory on why it was so high. But he implied that whoever built it must have had their reasons. He was a sheep farmer's son; his passion was walling.

As we talked, Steve worked through a large pile of stones, fallen from the wall. He would stop, stoop, hold a slab, bounce it in his hands as though to test its weight and edges, and then advance on the wall. It was not quite a jigsaw puzzle. He did not reassemble them exactly as they had been, nor spend three minutes looking for the perfect place. A stone could go in a number of places; what mattered was the overall design and direction. He moved steadily, placing, it seemed, about three stones a minute on the top of the wall. But equally, he was not taking everything. I offered what I thought was a plausible stone – a neat, thin rectangle.

'That's a bit too tall,' Steve muttered. 'Might go in. Will have to go in like that.' He placed it narrow side out, so that it stood on its edge and stretched halfway through the wall. 'I don't like doing that. It's called a soldier, that one, when they're stood up straight.' 'And you don't like them?' 'Not really, no.'

'What's wrong with them?' 'It doesn't quite look right to me, but anyway . . .' He laughed. 'How much can you build in a day?' 'A good dry-stone waller can build four metres in a day. A bad waller can build six.'

In Swindale, the dry-stone walls are now little more than abstract sculptures. This was always a remote parish: there was once a famous disagreement when the church bells rang about whether it was Sunday; finally resolved by the vicar's naked assertion of authority. But in 1580, there were eighty people in the valley, a church and a school. There is now no church, no school and no sheep or people to be seen. The valley is owned by United Utilities, the water company, which is in turn, it seems, owned by a hedge-fund in New York. The tenant is now the Royal Society for the Protection of Birds.

I was shown the adjoining land by officers from United Utilities and the RSPB. Both men were in neat hiking gear: fit, smiling; I might have taken them for canoe instructors. They wanted to show me how they had managed the land in a way that 'increased biodiversity, decreased flooding, increased carbon capture'. 'We don't believe,' one said, 'that there is any contradiction between good environmental practice and farming. We believe in including farmers, and retaining sheep.'

They led me into a side-valley: rushes grew deep and tall along the bottom; on the slopes were great fields of bracken, and beyond them scrub trees and, higher, heather. Pollen samples suggested that Bronze Age people had cleared this land and kept livestock on it. The new approach seemed to be restoring a landscape which had not existed almost since the first human settlement. I could not see any sheep. They told me, however, that they had kept about one sheep for every four or five hectares. I wondered whether officials in London who found this a reassuring sign of compromise with farmers really understood that this meant a single sheep lost in an area of thick scrubland the size of four football pitches. They showed me a section of eight hundred hectares which was being planted as forestry. 'How many sheep have you kept in that area?' I asked. 'None.' 'Are there areas in your plans, or the plans of the Lake District National Park, which have been designated to be preserved for dense sheep-grazing?' 'Not as such. But we have nothing in principle against sheep-farming.'

They employed a contractor, on a one-year annual contract, to look after their sheep. I suggested that densely cropped green lawns alongside the wilder fells had been for almost 2,000 years one of the beauties of the Lake District. A man on a one-year contract was not the same as a small family sheep farm with generations of occupation and security of tenure. Small family farms were links to the past: the last traces of our indigenous population. I argued that if the Lake District became a wilderness reserve occupied by professionals from elsewhere, we would have lost something very precious. And, rather than bringing professionals in from other parts of Britain, we should be running training courses for local farmers to do the same job.

We talked at cross-purposes. They replied by talking again about biodiversity, water management, sustainability and carbon capture. Farming for them seemed to be about employment, incomes, subsidies and environmental impact assessments. I sensed that behind this dry language they had strong views on what they thought was right, and beautiful. I guessed that they loved the idea of a much wilder landscape, more packed with wet bogs and bird populations; that they felt that farming and dense sheep-stocking were often destructive. They were careful not to say these things. But they did not seem comfortable engaging in a discussion about the history of the valley, about the traditions of small farms, or about the beauty of farmed land.

Swindale has always felt like a hidden miracle, a tight-necked valley opening into bright green fields by the river, a great tongue of land in the centre, neat and fierce as a fairy-castle, and the cold waterfalls coming down from pool to pool. In the bright sun, it might appear a corner of Jurassic Park.

But on the facing slope you can see the long lines of dry-stone wall running almost vertically up the fellside, enclosing the mountain-face in irregular geometrical slices. Each stretch is perhaps the work of a different generation: brothers dividing their patrimony and extending their grandfather's work.

In New Hampshire I have seen such walls only as broken hints, buried under a wilderness of scrub and forestry.

This piece made the RSPB officer who showed me around the project angry. He felt that having been warm to him at the time, I had betrayed him by mocking him and his project in this piece. He especially disliked being described as someone dressed like a canoeing instructor.

25 October 2014

Fell. Mire. Heath. Moor. Each generation has found its own particular upland landscape. On my walk last week, the rain didn't stop. I slowed to two miles an hour and stared at the wet ground. The colours were more extreme than I had remembered: the cherry-brown grass had turned scarlet, and the yellow-brown moss blazed saffron. The raindrops clustered in their thousands on every knot of heather. The orbs of glittering light pulled slowly away from their dark centres, so that each drop stretched into a crown and a pendant, like an acorn on its stalk.

I noticed the dense clusters of pellets which marked the nest of the mother grouse, and another brown grouse paste which I did not understand. A hundred yards further on, all a hawk had left was two black feathered wings, joined by a fragile skeleton. When I picked it up, I felt a sudden nip on my fingers from the beak in the bare skull, swinging in on its long spine.

I spotted a spider on a rain-flecked web. I saw the geese turning back from the low cloud, and the ridge-line, that interrupted their flight to Africa. An hour later, what remained of a baby hare suggested that it had just been on the point of growing its white winter coat. At the same time I missed and misunderstood a lot. I glimpsed an unfamiliar fern in passing, and regretted I had not stopped because I did not see another like it in the next six miles. What I had thought was a hind was only a thick tuft of yellowing grass, but what seemed to be twigs were in fact antlers. The stag rose from the hollow, only sixty yards away; it paused, sluggish after the rut, then, staring straight at me, shook the brilliant pearls of rain slowly from its sable ruff.

But the moorland was not as 'untouched' as it seemed. You could see subtle traces of human effort everywhere. Over thousands of years, farmers had often burned back the scrub and grazed cattle and sheep on this land. The patches of untidy grass and neat lawn were now hidden in heather, but the stones of the farmers' buildings remained.

Dorothy Wordsworth recorded the process in 1805: 'We were passing, without notice, a heap of scattered stones round which was a belt of green grass – green, and as it seemed rich, where all else was either poor heather and coarse grass, or unprofitable rushes and spongy moss . . . the heap of stones had been a hut where a family was then living, who had their winter habitation in the valley.'

A slate, marked with paint, showed that the heather bank had once been a butt for shooting grouse. A track had been a drove road, and the ruined building, a thousand feet below, had been built by a medieval bishop as a shelter for travellers. You could see the straight lines that marked the patch in the field where the house had cut its peat.

A thousand years ago such moorland was, for the *Beowulf* writer, a place for a demon. It was the lonely 'fastness . . . by misty crags . . . windy headlands, fenways fearful' in which the monster Grendel lived. Four hundred years ago, Shakespeare's King Lear found the heath a landscape in which to go mad. Charlotte Brontë made it a place for Heathcliff's despotism.

Dr Johnson found it offensively unproductive: the 'eye is astonished and repelled by this wide extent of hopeless sterility . . . matter incapable of form or usefulness . . . quickened only with one sullen power of useless vegetation.' Only a generation later, however, Walter Scott found it morally impressive: 'I like the very nakedness of the land; it has something bold, and stern, and solitary about it.'

Today, people in fleeces with the acronyms of different agencies on their chest can look at the same moorland and see in it not monsters, nor insanity, nor Johnson's sterility, nor Scott's virtue, but instead an engine for carbon capture, biodiversity and flood control.

For me, however, the fell seemed above all improbably alien. It resembled a tropical swamp seen from far above: the peat-puddles, lagoons; the heather, a mangrove swamp; the patches of pale, dying grass, a great savannah. The peat-hags on the facing slopes had the silhouettes of pot-bellied Mesopotamian gods. Nearer, it was a living reef. The lichens and ferns were coral, and the pink-brown moss a jellyfish. I pulled a piece of blackened wood from a bog. It was, it seemed, a piece of Scots pine, preserved in the damp soil from a time perhaps 2,000 years ago when this had all been forest. It had acquired the shape of a fish.

Each generation finds in this wet upland a different landscape. How will our descendants see it? What aspects of the future will repel them or involve them in this moorland? Certainly, their archaeologists and historians will uncover ever more about the ghosts of farmers who were driven from the empty slopes; their scientists will describe, over trillions of pages, every moss and lichen, grass and fungus; they will destroy and protect new sections of the heath, and they will discover fresh interests and obsessions which have nothing to do with the traditional British landscape.

Yet our great-grandchildren – modern, intimidated, disappointed and exhilarated – will, we hope, still enter the wet, treeless upland ground, thread a path through the heather, and conjure their own Grendels on the fellside.

8 June 2013

In the Middle Ages, most of the land of this constituency was turned into a 'royal forest'. Inglewood Forest alone between Penrith and Carlisle was sixty miles in circumference (if my maths is right, 286 square miles). It touched the Forest of Allerdale. If one adds Nicholforest in the north, and Gilsland, Geltsdale, Greystoke and Skiddaw/Thornthwaite, then more than half this entire constituency was a protected forest.

This did not always mean trees: it included swamp and heath. But where there were trees, it was not the carpet of Kershope and Kielder's dark Alaskan Sitka spruce, but ash, hazel, hawthorn, field maple and northern sessile oak. Forest meant not trees but a protected landscape, in which humans lived without normal legal rights, under the most extreme environmental restrictions and regulations. The purpose of course was to preserve deer for the king's hunting, but this was achieved through what we would now call rewilding. Forest comes from the old French word for 'something outside'.

For four hundred years and more, building any house or ploughing a field was illegal in half of this constituency: buildings could be destroyed on sight. The trees were protected and could not be felled; the grass was protected and could not be grazed; the land could not be drained, fenced or improved in any way; the turf and the peat could not be extracted. Everything was to be kept as a wilderness, to preserve the ideal habitat for red deer and roe deer and fallow deer; boar, hare and wolf; fox, rabbits, pheasant and partridge.

A bureaucracy of wardens and deputies, constables, foresters and under-foresters, revenue collectors, surveyors, judicial inspectors and rangers was employed to protect the animals and plants. (In some ways – if not in the hunting – it prefigures the legal framework, purpose and bureaucracy of the Lake District National Park.)

It worked. Some 250 years after Inglewood was established, the deer herds were so immense that the king could kill more than two hundred fallow deer in a single day. The impact of these restrictions can still be seen today, a thousand years later. Look left as you drive down the M6 to Penrith, for example, and you see the rich, green land on the west bank of the Eden, between Lazonby and Armathwaite, which you expect to be densely settled with villages. But, in fact, as I found two years ago, if you get yourself in trouble on the west bank of the river, you are in a seven-mile stretch with barely a road or dwelling. This was the part of Inglewood Forest which was marked on sixteenth-century maps as 'barren park waste'.

The problem with the old forests was not the objective of protecting the environment, but the rigidity with which it was done, and lack of space for humans. The cost was to the lives of communities. It was a good place for outlaws like Robin Hood, Adam Bell or William of Cloudesley. But farming was so restricted that almost no income could be made from the land.

The only employment was in being paid by the government to protect the landscape or in supporting the rich people from other parts of Britain who came to enjoy it. We know almost nothing about the communities that tried to survive in this context, since they lived in the shadow of the law. What we know about the distinctive culture and the life of this constituency was then found almost entirely outside the forest, in the small farms of the Lake District hills, with their tiny chapels, Norse-influenced dialect, unique funeral customs, and thousands of small, independent, owner-occupier statesmen.

The farmers who lived on the edge of the protected land were understandably furious. King after king, from Magna Carta onwards, promised to reform the system. And finally, the reforms came. The powers of the rangers and the managers were reduced in the late Middle Ages; locals were granted limited rights to build small properties, to graze and cultivate and harvest; local figures became more closely involved in conservation and management.

But the reforms were too slow, and pent-up frustration ultimately led not to a more reasonable compromise between humans and nature but to a complete reversal: Henry VIII tore up the entire framework that had once protected the environment. In the free-for-all, the oaks were felled, many for props or fuel in the new iron mines in Cumbria. Within thirty years of the death of Henry VIII, the map of 1576 shows only a belt of woodland near Carlisle – the rest of the trees in Inglewood had gone. By 1630 John Aubrey describes an almost bare England, facing a crisis of timber. The wild boar and the wolves had been long killed; the red deer moved north, and the fallow south, and the martens went with their habitat. You can see the trace of what followed in aerial photographs – the centre of our constituency, where the forest once ran, is defined not by the curving lines of the old Celtic fields, or the corduroy stripes of the Anglo-Saxon common fields, but a Kansas-straight grid of early modern agriculture, neat as AstroTurf.

The habitats which our ancestors tried to protect with forest law could not be sustained because they failed to find a balance between the habitats and interests of the animals, the recreation of visitors, the income for farmers, and homes for communities. This is still the challenge in national parks today. Such a balance can be found, but it cannot be achieved simply through targets and legislation. It requires intelligent pluralistic planning, rooted in knowledge of particular landscapes and local families, in which one Lake District valley might be designated for wilderness, but another protected for traditional sheep farming, or in which a textured balance is achieved in a single valley – where 'nature-based farming' is not simply a euphemism for business as usual, and 'lighter stocking density' is not simply a euphemism for removing sheep farms. In the long term, the environment can only be protected if we understand the needs of our most peculiar ancient native species, the human.

If we fail in the balancing act, we will not be able to protect humans or the environment. We will be left with what Camden saw when he visited Inglewood Forest in about 1600, 'a dreary moor with high distant hills on both sides, and a few stone farmhouses and cottages on the road side'.

17.30 from Euston, 27 October 2012

Coach E on the evening train home. And one row in front of me is Shoshana: soon to be my wife. I am worrying whether I have the correct arguments to save the defence munitions site in Longtown; and wondering what the minister thought of the briefing on Eden District Council. The middle-aged couple next to me have loud, cheerful voices, which makes it tricky to concentrate. I must ring a Cumbrian landowner to convince him not to charge 'way-leaves' for a community broadband project. There are more emails than I like to think about in my inbox, a half-eaten BLT sandwich at my feet. We are getting married on Monday.

We are getting married on Monday: I repeat it to try to understand it. On Monday at 11 a.m., fourteen of us will stand in the small, low chapel in Scotland – our parents, her brother, my sisters, my nephew: everyone from home, and the vicar. And that is basically it. Except, perhaps, my father's lurcher Torquil, who may or may not appear in a red tartan bow-tie.

Shoshana and I have made the programmes. We found Japanese paper from a bookbinder; I typed out the order of service; she printed it on our home printer and bound the leaves in sea-grass thread. We will go to and from the church in the family's Vauxhall van, which my nephew should be decorating.

And we'll walk the last section home over the fields, which should require wellington boots.

It is unclear what the Aberdeen Angus and the Highland cow in the sunken field will make of her wedding dress. Or indeed what I will make of it, since the dress remains a secret – though I can see its shape suspended in a long bag from the hook beside seat 32D.

I feared that my friends who have spent the last few years among babies and children would see my stag night as a rare chance to relive their lost roaring youth. They sent a stream of emails. One contained the sentence 'this is the dress-code' with a link to a YouTube video of the Woodabe, a nomadic African tribe who appear to dress in women's clothing. They said they would turn up at seven in the morning, and implied I should have a passport so I could get home if they abandoned me on the ferry to Zeebrugge. Shoshana kindly went to stay with my aunt for the night.

But my friends, like me, are middle-aged. Seven o'clock became 'I will call you in the morning'. The organiser turned up first, at three in the afternoon, with a bad back, and lay on the floor giving me a lecture on economics. Silver-haired fathers arrived whom I had first met thirty years earlier, when they were three feet six inches tall and bouncing on trainers fastened with Velcro. They bounce less now, but their five-verse song mocking my political career proved that they could still sing: quite loudly enough, I thought, for the other diners.

They mostly turned in before midnight, heading for late trains and the babysitter, leaving me with scraps of advice. Some seemed promising ('She needs to start with very low expecta-tions'); others were confusing (Luke: 'Set boundaries early: never do "night feeds".' Me: 'Is that what you did?' Luke: 'Not at all'); others, terrifying ('It will go wrong, and it will be all your fault, and it will be your sole responsibility to mend it').

I can see her through the gap in the train's red seats. Shoshana is watching a film on her computer with her headphones in. She is smiling and shaking her head: sometimes frowning. It looks like she is concentrating on a biology lecture, but she's watching *Star Trek*. I'm surprised she's awake. She hasn't slept a full night since she got back from Kabul a week ago. She has been corresponding today with a charity on inner-city education, organising an exhibition, showing her brother round London, dealing with a difficult and bitter colleague, and replying to every email from her mother and my mother. ('Is it okay if your nephew hitches a ride in the honeymoon car to the airport with you?') She is wearing a dark blue silk jacket she bought last year when she was studying in China. Her hair is held back in a bundled mass of brown curls, tinged with red in the train carriage lights.

On my finger is a piece of bruised and flattened gold, which she has given me. It is for the moment on my right hand. It has a cross in a circle on the top, and is from Byzantium. I guess from the size it's a man's. Was it only one man's? Or was it passed on for generations? Was it admired by his wife? It's from what is now the Middle East, where we two first met. But it is also as Roman as Hadrian's Wall, and as medieval as Rose Castle, and the cross is the shape of the Bewcastle Cross.

I am in a rumpled suit, tie at half-mast, typing on the evening train. But in the reflection of that black window, I can see her white teeth smiling. And I can see the gold ring, which I gave her, on her right hand: also waiting until Monday for transfer to the left.

HOMELAND

I SPENT TEN YEARS INSISTING THAT what mattered in politics was the most local and concrete engagement with a particular place. I ended my political career over that more abstract and continental issue, Brexit. A slim majority of my constituents had voted for Brexit. I voted Remain, partly because I felt the European Union's tariffs and subsidies were the best protection for small Cumbrian farmers – but I accepted the result of the referendum.

I campaigned for Theresa May's softer Brexit deal, which would have controlled immigration while remaining in the customs union. I failed in that too, and when a hard Brexit followed in 2019, I resigned. Single farm payments ceased and were replaced by a contract to improve environmental outcomes. Whereas the single farm payments had been automatic, these new grants required individual Cumbrian farmers on very low incomes to complete complex grant applications and enter expensive monitoring schemes. The grants were competitive: not every farmer qualified, and the large, wealthy estates, and charities such as the National Trust and RSPB, found it much easier to employ land agents to construct proposals for the more sophisticated schemes. When Labour was elected in 2024, and faced a rising deficit and an urban voting base, they closed these government grants before the upland farmers had a chance to access them.

Small-scale Cumbrian farming will now I believe become

close to impossible. The large flocks of Herdwick and Rough Fell sheep will fade away, and with them the emerald colours of the in-bye; the dry-stone walls will eventually crumble, wilderness will encroach across the *ring-garth*, spreading from the fellside down into the old pasture and meadows. The remaining farmhouses will be sold to wealthy incomers as holiday or retirement homes. What remains of rural tradition and identity will be leached away: all that was cussed, charming or distinctive replaced by wilderness and forestry. In the more prosperous areas, there will be neat lawns, and gastro-pubs for commuters and tourists. But most of the remoter valleys will follow the patterns I had seen in Bewcastle: depopulation and the collapse in quick succession of auction mart, shop, church, pub and school.

I resigned from the Conservative party when Boris Johnson tried to railroad his particularly damaging version of Brexit through Parliament. And since I did not want to run as an independent against the team of people who had supported me for ten years, I also resigned from Parliament in November 2019. In 2023, the Boundary Commission abolished the constituency itself. Penrith and the Border ceased to exist.

I end with pieces drawn from across my decade in Penrith and the Border. I am struck again by how little connection this collection of letters has to modern politics. Their conservative attachment to history, to tradition, to a slowly evolving, human-shaped landscape and to small family farms has no space in the modern Conservative party. Let alone any other party. They do not promise salvation in the form of AI and quantum computing. Or even Green Growth. And they seem opposed to many of the ideas of economic policy, reform or renewal, which are at the heart of the US debate. (I suspect that, unlike me, Joe Biden would have backed Jack Adams' industrial policies. Elon Musk would have applauded the radical efficiency and technological prowess of the Roman occupiers, and Donald Trump would have admired the rejection of law and international cooperation epitomised by the medieval warden of the Western March.)

The consistent theme of these letters is the way that central schemes designed in London – let alone grander global visions – fail, precisely because of inability to acknowledge, or understand, an organism like Penrith and the Border – in which 20,000 tiny businesses, 2,000 farms and thousands of community institutions and services sustain and feed each other. Each on its own is fragile and easily destroyed by new laws or budget cuts; each destruction ripples through the valleys in unanticipated ways.

But this is not an argument for freezing Cumbria in the past. Rather, I feel that we need a new Northumbrian renaissance – a period of intense experimentation and importing of ideas from other countries. We need to be open to population movement as Cumbria was historically with the Norse Vikings and the French monks. Much of the greatest vigour and community energy in Cumbria continues – as it always did – to come from people like Mary Burkett, who was not born in Cumbria and spent a great deal of time outside the UK.

Nor should we defend bad institutions. We shouldn't be protecting poor farming practices: over-stocking, slurry-spreading and nitrogen release into rivers, or inhumane animal husbandry. Instead, we should be learning from individuals like the Herdwick farmer, my constituent James Rebanks, who combines a fierce traditional attachment to family farming and rare breeds of sheep. He has proved that better environmental outcomes can come from smart grazing, rather than from stopping grazing entirely. Improving lives requires building new institutions. Instead of resisting the Lake District National Park, we should be using its UNESCO World Heritage status to embed the importance of traditional farming and to frustrate the more extreme ambitions of the large estates and some of the environmental charities.

Defending traditional culture requires not constitutional conservatism, but a revolution in our political system, and learning from radically different traditions. Cumbrians are the victims, not the beneficiaries of our tribal politics in Westminster. There needs to be a radical devolution of power, not simply down to a 'Mayor of Cumbria' level but (learning from France) down to the level of a Penrith or Wigton mayor with real financial power and authority. We should introduce Australian compulsory voting and a New Zealand electoral system to force politicians to persuade the centre rather than indulge the extremes.

But most of all, I hope all the letters in this book show – in part by illustrating my flaws and prejudices as an MP – why our current model of representative democracy is broken. Impossibly, irretrievably broken – and not fixable by just electing a better MP. These essays are a continual hymn to the power of Cumbrian communities. From upland farming to community housing, Cumbrians demonstrated to me over a decade that they consistently knew more, cared more and could do more than distant officials. But they were not given the political space to exercise their wisdom.

These letters are, therefore, also an argument for reforming a political system with citizens' assemblies at its heart. Learning here not from ancient Cumbria, but from ancient Athens (with perhaps a small debt to the Norse thing-mounts of Threlkeld). Citizens should be selected randomly in accordance with demographic data (as we do with juries) to investigate and discuss issues, under the guidance of experts, who ultimately leave the decision to the citizens.

At a time when we are tempted to think the answer to our struggling democracies is to become less democratic – to consult less, and become more technocratic, more growth-focused, more decisive, more as it were like my father's dream of a colonial district commissioner – I disagree. I suspect the answer to a national crisis in democracy is to become more local and more democratic.

28 September 2013

Two hundred years after the Romans left in AD 600, some Cumbrians must have felt their world had ended. Agriculture had collapsed around them; the population had plummeted. No stone building or new roads had been constructed in two centuries. Education, industry and trade had collapsed. We were one of the most underdeveloped places in Europe or Asia.

But within just two generations our remote, sparsely populated area was producing the greatest art, spirituality and scholarship in Europe, partly because our rural isolation became a strength, not a weakness. We were transformed, first, by a new faith. Christianity arrived in heathen Northumbria and Cumbria in two ways: with charismatic Irish ascetics, travelling on foot; and with horse-borne bishops sent by the papacy. We were ideally placed to combine these rival traditions because we were always a Middleland. When Hadrian's Wall was manned, we were half part of Rome, half outside it. We were never part of Roman urban civilisation: our landscape and culture were more like 'barbarian' Ireland. But we were surrounded by the great walls and forts of Rome, and had touched a wider European civilisation.

Then, we were transformed by our curiosity. We sent our own scholars on the impossibly long journey to Rome, and eagerly copied down all the knowledge with which they returned. When a Syrian missionary arrived in the eighth century, our scholars assailed him with so many eager questions that a witness compared him to an old boar fending off a pack of puppies. We sent monks from northern England to learn from the best musicians, masons, glaziers and scholars on the continent. We studied crisp carving and orthodox images from foreign sculptors. Then we surpassed them.

On the Bewcastle Cross, for example, Cumbrians worked a sundial across a petal, invented unprecedented flowers, and filled an entire frame with a mystical chequerboard. But the dignity of the figures and proportions of the composition remained in the best classical tradition.

The culture of eighth-century Cumbria and Northumbria was not static, insular or parochial. It depended on our capacity to use with confidence the energy of different traditions. We preserved some of the tone of our own pagan past, and we retained some of the purity and spirituality of our Celtic Cumbrian Christianity. Hints of this approach survive in our unique Cumbrian saints such as St Ninian and St Kentigern, who took their distinctive, very British message from this constituency up to the Picts in the Strathearn valley of Perthshire. We abandoned the most discredited customs of Celtic Christianity and embraced some of the latest ideas from Rome, but we lived ascetic lives, which world-weary Romans had thought no longer possible. Within forty years, as the Mediterranean declined, Northumbria and Cumbria were producing the greatest artists, scholars, missionaries and statesmen in northern Europe. Bede, the greatest historian of his age and one of the finest late writers of Latin prose, came from a culture which had been, not long before his birth, almost illiterate.

It is not Bede, however, nor St Ninian, or even that

Cumbrian-born missionary, St Patrick, who best embodies our culture at its most distinctive and vigorous, but St Cuthbert. He was the ultimate symbol of our Middleland civilisation because he was simultaneously profoundly embedded in our most local landscape and communities, while retaining the broader ambitions and experience of an off-comer. He was born in what we now call Scotland, and died in what we now call England, but his life and practice often echoed an older, native Celtic Cumbrian sensibility. He retained an almost pagan delight in animals – he was fed by sea-eagles, and communed with ravens. According to an eye-witness, he stood all night up to his neck in the sea to pray, and at dawn, otters came to lick the frozen saint back to life. He combined this, however, with a great reverence for scholarship, acknowledged that he was part of a broader European civilisation, and died as an orthodox bishop, encouraging his disciples to follow the customs of Rome. He died on a sea-girt island, an Anglo-Saxon monk in the Roman tradition, suffering alone as a Celtic ascetic. And it was because of men like Cuthbert that the Pope, looking for a missionary to the Vikings, turned to northern England, and that Charlemagne, looking for a chief of staff, chose a Northumbrian.

Our golden age has never been easy to admire, or even remember. It left no Ziggurat of Ur, no Machu Picchu or pyramid. Many of its most distinctive contributions lay in advances in religion and theology, which we struggle to understand. Even its most famous treasure, the illuminated pages of the Lindisfarne Gospel, is not a public monument; it is handwritten in an alien language. You can't admire its pictures side by side, as you might in a modern art gallery because it is the only copy of the book. The turn of each page hides the last picture as it reveals the next. All that survives of the seventh-century Hexham Abbey, once the greatest building of its kind north of the Alps, is a narrow crypt made of grey-stone lifted from Hadrian's Wall. Of the major Anglian monastery at Dacre, fit to be visited by kings, no trace remains beneath the stone beasts in the churchyard.

Yet, no other civilisation has come so quickly, from rural isolation, to dominate the imagination of a continent. None has made such unpromising conditions a more rapid catalyst for seriousness, and greatness. It was a golden age lived to its fullest in places not just without cities, but without buildings: in the red sandstone cliff walls of the Eden at Wetheral, or on the island in the lake at Derwentwater. At Lindisfarne it is easy to be transfixed by the ruined priory, with its purple columns tapering like sandstone pillars scoured by desert winds. But that building was constructed centuries later. The real essence of the northern renaissance lies further out to sea, in the faint shape of Inner Farne: a place defined by the iridescence of the water at first light, by seals, and by birds. St Cuthbert's final home.

4 September 2010

I have spent the last two days walking alongside the Eden. I thought I would experience and remember it as a single flowing stream. But instead it seems to be many different rivers. At Mallerstang below the falls the valley is narrow, with folds and rivulets and short limestone distances that conceal castles. But the dark, rearing crest of Wild Boar Fell seems part of an entirely different landscape, looking over all 1,200 square miles of the constituency from Blencathra to Bewcastle.

After Kirkby Stephen the river is tranquil and measured in its meanderings, at odds with the framing line of the Pennines and the volcanic pikes behind. A sudden wall of pink sandstone at Temple Sowerby seems the setting of an oriental mystery; but two miles later, approaching Langwathby, the flood plain is bare and treeless, and the gravel scattered with desultory cows. Tonight, before Lazonby, I am sleeping near dark rock-falls and caves.

Eden has given its name to a district. Many of our community groups take their names from the river and its tributaries. It is the water that some of us drink, a landscape we love, a lure for tourists: an artery. But walking along it, I sometimes feel we have abandoned it. From the banks one glimpses the backs of houses, sewage treatment plants, dead-end tracks, slurry tanks.

In Hartley, or Morland, the houses look towards the stream and the millrace. But those are unusual tributaries. The main river often flows behind and apart from the villages. Even accessing the banks feels furtive: that one is discovering something almost like a disused canal. In places it is only half-accessible, down crumbling sandy cliffs, ringed with old barbed wire and touched in places by Japanese knotweed.

Today I was joined on the walk by a grassland expert, an organic dairy farmer, a professor of soil science, a water habitat expert and a man who had fished the river for fifty years. They showed me how sometimes, every mile, the nature of the river changed completely because of a new stone-base or even the barrier formed by a bridge. They let me handle white-clawed crayfish and told me the fish live only in limestone because they need the calcium for their shell. They explained how the Borrowdale volcanic rock in the River Derwent meant water poor in nutrients and filled with oxygen, delighting certain flies who avoided the nutrient-rich Eden. They showed the riffles, which pleased the new-hatched fry, and the pools for the parr and the runs for the adult salmon. They explained the impact of fertiliser, stock and soil management. They showed how phosphates had encouraged the algae, which choked the gills of the crayfish and left the three types of lamprey unable to breathe above the silt.

But when I heard more about what this meant for farmers – the £65,000 tanks that need to be built for slurry to keep nitrogen out of the river, the banning of bare fields on the water edge, the emphasis on core sampling and soil analysis, the specific times of year in which fertiliser should be laid, the strict controls on stocking – it looked much less easy. How, when margins are so tight, can farmers be expected to invest so much money and time and research in something whose effect on their farm is often, at least in the short term, very indirect?

What impresses me most therefore about the Eden Rivers Trust is the effort it makes to engage with farmers. Robert Warburton, the chairman, is a farmer; Will and Tom, from the trust, who accompanied me on the walk, are farmers. Hundreds of volunteers are working with the trust. I came across them on Wednesday counting trout and on Thursday counting crayfish. They remove abandoned objects from the river near Carlisle, and they weed Himalayan balsam from the banks. Most importantly, they get things done.

At Hoff, for example, I saw a ford which had blocked the movement of fish. The agencies had complained about it for decades. But the ford was the only way of getting the milk wagons into the dairy farm, and new work carried many risks for the river. So the agencies had done nothing. It was the Eden Rivers Trust that ultimately convinced the farmer, reassured all the agencies, found the contractor, filled out all the paperwork and had the culvert installed in ten days, often working at night without disrupting the milk truck. The fish are now spawning freely again.

My previous walks have taken me over, away from, or high above the river. I have experienced the landscape largely in terms of slopes and trees, boulders and peat. Suddenly I am able to hear its movement. I begin to mark the different geologies, and understand something about the algae and fly, fish and fertiliser. I am learning how (often for very understandable reasons) we have polluted or tried to tame the flow. But what I will remember most are the farmers and volunteers who are daily protecting and preserving this, our river: our Eden.

4 June 2016

Shortly before my father died, he was reading *The Road to Wigan Pier*. He had been struck by Orwell's description of the industrial scars and the poisonous detritus in northern England in the 1930s. I had been walking a great deal, and my father wanted to know whether I too had found a landscape 'wrecked by too many centuries of industry; too many people in too small an island'.

But in fact, our environment is, in many ways, in a much healthier place than when Orwell was alive and my father was young. Acid rain has been removed from the air, and many chemicals from the rivers. Some of our beaches have never been so clean. Our engines have never been less polluting. Mile after mile of new rowan, birch, oak and ash have been planted by highways, on old rail tracks and along the sides of deep gills. The increase in woodland cover is staggering. When my father was born, only 2 per cent of England had been covered with woodland. The figure today is closer to 10 per cent.

Part of this is due to economic change. The lead, iron and bauxite mines of the Lake District, and the coal mines of the Cumbrian west coast, which employed thousands in my father's youth, are all gone. So, too, are almost all the military munitions sites, airfields, nuclear plants and secret test sites which were scattered across the Border valleys forty years ago.

Part of it is about different approaches to agriculture. The most extreme over-stocking of sheep disappeared with the old headage payments. And we are getting much better at agri-environmental schemes. Walking through Peppering Farm in Sussex, for example, three weeks ago, I saw how by moving to smaller fields, wider hedgerow margins, more crop rotations and more feed for birds, the Arundel estate has combined profitable crops with an explosion of birds: skylark numbers have tripled, lapwings doubled, and grey partridge have increased almost one hundred-fold since 2003.

In Knepp, in mid-Sussex, I have never heard birdsong so loud and beautiful, or been more moved by the sight of Tamworth pigs, long-horn cattle and red deer moving through a primeval savannah of unruly flowering hawthorn and thick willow. This is a tribute to the work of two people, Charles Burrell and Isabella Tree.

In the National Forest, the secret has been a DEFRA-funded scheme of tree-planting which, over twenty-five years, has turned the bare mines and potteries of the West Midlands into a paradise of glades and copses.

And last weekend I saw how the Peak District National Park has restored bare black peat, poisoned by two centuries of acid rain, back into native grassland, heather and sphagnum moss.

We are able to count improvements in numbers of farmland birds, tonnes of carbon, or cost of production. We can already see the impact of recent changes in much of the English landscape: far more birds of prey in the sky, more badger setts and more otter holts. (But also, fewer farmland birds, fewer wildflower meadows, fewer hedgehogs or water voles, and certainly fewer salmon). All this change reflects not only different impacts of pollution and population pressure but also the ability of the different large estates and environmental charities to bend subsidies and regulations to their different agendas.

But where does this leave the Cumbrian landscape? Our small, traditional sheep farms often represent exactly what some interests think they want to eradicate. The environmental movement is tempted to argue that the sheep are damaging to biodiversity or water quality, and agricultural economists are tempted to argue that small farming is inefficient and lacks 'economies of scale'. They rarely acknowledge that this Cumbrian landscape of small farms – where every slope, barn and flock carries the trace of a thousand years of human cultivation management – is unique in Europe, and uniquely beautiful.

So, how do you promote projects such as the farms at Peppering, the rewilding at Knepp, the reseeding of the Dark Peak, while still maintaining Lake District hill farms? The answer, I believe, is to emphasise something which is not easy to measure or justify in a government document: beauty. This beauty is not without practical value. It is beauty that ensures that millions of people pay to visit our landscape, it is beauty that will guarantee that future generations care about our land and defend it. And what has struck me in my recent walks is that these environmental and agricultural projects, including the rewilding in Sussex, are obviously and profoundly beautiful.

Beauty is a more generous, open idea than slogans about food production or the environment because it leaves room for the human, the historical; for a dry-stone wall or a medieval bank barn, as much as a productive crop; for a Herdwick lamb as much as a lapwing. What I should have said to my father is that if his generation's problem was too much human interference in the landscape, ours is sometimes the reverse: we risk losing our belief in a positive role for humans in our landscape, and the way to restore it is to be less embarrassed about admitting what we love.

1 August 2015

Of the thirty barns and houses I can see scattered along the valley beneath my cottage, only one was built after 1800. The sun, striking the facing slope, picks out three colours: brown for the ploughed fields; olive green for the hay meadows; and a yellow-green for the enclosed sward of the sheep pasture.

The glittering limestone walls which cut these colours into irregular shards, framed by the russet band of the high fell, are old; and it is sometimes tempting to feel that someone, at my cottage, would have seen the same view for a thousand years.

And yet, in truth, this land is more changeable than the surface of the sea. The place-names record the very different nations that once lived in this valley. The name of my cottage, Cragg, is an old Celtic word; the village to the north-east, Brougham, is named after a first-century Roman fort; the valley to the east, Lyvennet (a lime grove) is a name from sixth-century Welsh-speaking Rheged; the valley to the south, Swindale, is ninth-century Norse for a pig-pasture; the village to the north, Yanwath, is tenth-century Strathclyde-Cumbrian, combining a Celtic and a Norse word for a single ford.

The names on the rich riverine soil in the heart of this valley – Helton, Bampton, Askham – are Northumbrian names, from the seventh- and eighth-century Anglian settlers. And all these nations will have used the land in different ways. The Celtic field-shapes would have been circular, the Northumbrian, narrow strips; the Norse focused on upland pasture.

The soil, however, holds records of more than just these different ethnic and linguistic traditions. Pollen in the peat preserved the first felling of the forest in the Bronze Age by people whose language remains entirely unknown; a dig near Yanwath uncovered dry-stone walls, two millennia old. The evening light reveals the shadows of furrows on the upper slopes, which suggest that the appetite for wheat from 15,000 Roman soldiers drove ploughs into marginal land.

A circular wall is the remains of a thirteenth-century deer park; the deep gullies which I can see in the peat-hags were cut by the spades of medieval villagers; and the scars on the further hills were made by sixteenth-century quarrying and mining. Higher still, you can find the drains and the lime kilns of the nineteenth-century improvers, and the abandoned traces of an Edwardian grey partridge shoot. And at the ridges, Neolithic standing stones.

The evidence hints at retreat as well as progress. Subtler traces of pollen imply that after the Roman departure the population collapsed, and scrub and wild animals reached almost to the valley floor. Monastic records suggest that what is now a patchwork of fields would have been a single russet band of heather when it was a medieval forest in the twelfth century; then pale green, when the circular park-wall fenced the deer in the thirteenth; then bright emerald in the early fourteenth when the monks began to allow seasonal grazing. And it would have been russet again when the Black Death, the cattle murrains and the Scottish raids drastically reduced the population once more in the fourteenth century.

Nor is the modern population very ancient. I know one family which came here from Birmingham in the early nineteenth century, and another which saw Bonnie Prince Charlie's army advance along the ridge-line in 1745. But the majority of the cottages are now inhabited by people who were not born in the valley: retired schoolteachers from Manchester, a bishop and an MP born in Hong Kong. Only the Lowthers claim they were here at a time before people spoke Modern English, when there might still have been traces of Norse and Cumbrian Celtic, and when they spoke Norman French. And even the Lowthers perhaps were not here forever.

The families who first felled the trees, began the dry-stone walls, first ploughed, built the first house in these hamlets, first turned a wilderness into a pastoral paradise, are gone. We do not know what colour their hair was, or what language they spoke. We have all inherited a stranger's work. There is so much change that it is a shock to remember that even 10,000 years ago the river was still there on the valley floor.

I wonder if any previous generation, looking at this valley, saw such a shifting video of shape and colour, of cultivation swelling and receding, over six millennia? Or is this sense of continual change just the product of new advances in archaeology and palaeobotany? Or a reflection of our daily diet of time-lapse photography and computer-generated images?

Change accelerates. Thirty farms in this valley have become twelve. The short-horn have gone, dairy has gone. Nobody keeps a pig any more. The sheep numbers are dropping fast. Even over the last ten years, the fell has begun to change. A green sward has vanished, and a spiky hedge of gorse bushes blocks the route from the swing-bridge. Bracken has overtaken the upper slopes. Scrub-trees grow along the becks. The drainage ditch is no longer cleared. The river is not dredged and floods flow freely over the plain. On Helton Common, great beds of reeds replace the old pasture.

How will our successors judge our use of land? Praise us for protecting carbon in peat bogs? Admire the trees and the species which have flourished in the longer vegetation? Or regret the loss of productive land, the reliance on intensive farming, and the imports from other countries who are less careful with their environment?

And who will decide what the landscape looks like in twenty years' time or two hundred? An NGO? A scientist? A politician? The people who have retired recently to the valley? The remaining farmers? And if, in three hundred years' time, some descendant climbs Knipe Scar to look down on the valley, what will they see? What colours, what textures, what houses, what people? What will they recognise of our landscape, except for the river?

9 November 2019

At about eight yesterday evening, there was the sound of crying from upstairs. Was it the two-year-old? Had his rabbit fallen out of bed? It was the four-year-old. I picked him up. He was crying so much he made my shirt front wet. I couldn't understand what he was saying.

At last I got it: 'I don't want to say goodbye to the sheep house,' he said. We sat in an armchair and he told me about his love of Cumbria. We discussed the Lost Castle playground at Lowther ('Why is it lost, Daddy? Why did the wicked fairy send Sleeping Beauty to sleep? Why was she angry not to be invited to the christening? It's okay to forget to invite someone to a christening, isn't it? It's okay to forget sometimes?').

He grumbled about Boris Johnson. Some four-year-old's history lesson had persuaded him I am Cardinal Wolsey, being stripped of my golden seal of ministerial office, my seat in Parliament, and even my Cumbrian cottage by a modern Henry VIII. And then we discussed his Everest – Knipe Scar; and Swaledale and Rough Fell and Herdwick sheep. And I promised that we would still spend time in Cumbria. More time perhaps, now that I had left the Cabinet.

I fell in love with our county through its landscape. I learned the constituency by walking between and through every one of the 120 villages in Penrith and the Border. Nothing summed it up better than Bewcastle – tucked between Hadrian's Wall and the Scottish border. Once I walked fifty miles to get there, and a couple of times I clattered around on a horse. The hills are low, the fields rushy, the views limited.

At first sight you notice the farm sheds and little else. But look more closely and you can see that the mound is faced with great ashlar blocks that enclosed a pre-Roman Celtic shrine to our local god Cocidius. (He was a war god that looked like my four-year-old's picture of an alien, complete with antennae trembling on his head.)

The Roman fort that stood on that hill once housed a thousand soldiers from what is now Romania. When the Anglo-Saxons turned it into a royal sanctuary, they carved the most distinguished piece of stone carved in eighth-century Europe – doves and robes, tartan and sundial – clear, restrained, aristocratically confident. And then the stone was re-used in the great ruined castle – the keep of the Captain of Bewcastle – stronghold of the cockpit of the Border reivers for three hundred years.

But over the last decade, my love of Cumbria has shifted from history and landscape to living people. Libby – fierce, cheerful, certain to prevail, whether handling a pony or laying superfast broadband fibre in defiance of all paperwork, beside a mole plough-driver in Mallerstang.

David, whose subdued, stubborn energy created the beautiful affordable housing in Crosby Ravensworth. Malcolm, once a Scots Guardsman and police dog-handler, turned district councillor, passionately sympathetic to the needs of his ward, and with a fine tenor rendition of *The Pirates of Penzance*.

Steve, Trevor and Colin, working the heavy clay soils of Bewcastle – soaking soils, poached by the cattle's hooves – sustaining, against all the pressures from forestry and finance, their fierce attachment to the chain of small farms along the Bailey valley.

Richard and Cressida, Tom and Totty, Ilona, Henry, Philip, Antony and Elizabeth – warmest, most courteous hosts and friends, deep in knowledge and love of their land. Philip, with forty-one direct recorded generations before him, but who would not thank me for too many column inches on this page.

Dave, whose rugby player's build can withstand the Helm Wind when dragging a sheep's carcass from a snowdrift before returning to his delight in the *Guardian* newspaper.

Adrian, a tireless, exuberant campaigner in Tebay, impatient with government, and also an angler delicately sensitive to the chemistry of the Lune.

There is Wendy with her red squirrel kittens and continual kindness of jams and kale. The bishop still bent over his austere and lonely scholarship. Stead and Marilyn.

My friend Robert whose imagination expands every year with his farming – and a new understanding of grass, of food and of the management of rainfall.

Finally, there is John, my closest friend, impatiently patient, listening volubly, exactingly trusting; mentor on otter poo, and punctuation.

I leave politics troubled. As MPs, parties and Parliament, we have not governed well. It is twenty-nine years since I first entered public service, and during that time I have sensed the fading of our institutions – army, Foreign Office, civil service, the BBC and the rest. And yet, the people I have come to know in Cumbria have not diminished but flourished – in grace and common sense; in the hardihood I have seen in floodwater and snowdrift; in pride and courtesy.

Here, I found in dozens of individuals a wisdom, which – if we only had a democracy brave enough to harness it – could make us deeply proud, not just of Cumbria but of Britain.

Acknowledgements

Thank you first to Colin Maughan, the editor of the *Cumberland and Westmorland Herald*, who took a risk on me and gallantly published these articles week in, week out for almost ten years, and held me and all officials in the constituency fairly to account. British democracy would be in a healthier place with more Colin Maughans and more *Heralds*. Thank you also to his successor, Emily Atherton, who worked closely on this manuscript and has supported an eccentric project with conviction and vigour.

Thank you to Catherine Anderson, who took a bigger risk on me, ran my constituency office, bullied me to meet these deadlines, processed these articles and got them into the *Herald*, while managing so much prodigious positive activity at the same time; and to her team and all her successors, who took over her role and kept this going.

A thank you to my close friend John Hatt, to whom I have dedicated this book. He was sceptical about this project and frustrated with me (as usual) for doing things in too much of a hurry, so I must make it absolutely clear that he's not responsible for the finished product. He nevertheless dropped everything to help me get these pieces together, making – on his count – seven hundred suggestions. Of which I accepted 684.

Thank you to Clare Alexander, my agent, who has championed and represented me for twenty-five years – beginning with
an unpublished man in his mid-twenties about to go on a long
walk. To Bea Hemming, who for the second time has commis-
sioned one of my books and wisely, patiently and intelligently
guided me through the writing.

A thank you to my wife Shoshana, who took the biggest
of all risks on me – who listened to me reading and discussing
many of these letters when I first wrote them and balanced
supporting me in this with everything she does for the beneficiaries of Turquoise Mountain in five countries, and also
every day for all that is precious and valuable in our family. To
Ivo and Sasha, who put up with my absence weekend after
weekend as I struggled to bring this collection together – I
missed playing with you and I hope you will someday find
some of this project worthwhile.

And finally, a real thank you to my mother, who, as soon
as she heard the book was coming together, demanded to
be sent a copy. She then – aged eighty-nine – set to proofreading against a very tough deadline, putting in what must
have been eight or nine hours a day. It's a role she has played
so often for me. She will still be frustrated that there wasn't
more time for the kind of detailed proofreading in which
she excels. She will also regret that I did not emphasise more
what she experienced from this book – that it was, in her
words, 'an elegy to an ancient constituency which has now
been abolished and to a vision of politics that is under threat
everywhere'.

It is now ten years since my father died, and every year that
passes brings out more clearly my mother's strength of character, her grace and cheerfulness, the patience she has with our

family, her tolerance, her understanding, her sense of fun, her intelligence, and above all her confident, continual expressions of love. A thank you not only for what she did for the book, but also for everything that she's meant and continues to mean to me as her son.